NEWTON AS PHILOSOPHER

Newton's philosophical views are unique and uniquely difficult to categorize. In the course of a long career from the early 1670s until his death in 1727, he articulated profound responses to Cartesian natural philosophy and to the prevailing mechanical philosophy of his day. *Newton as Philosopher* presents Newton as an original and sophisticated contributor to natural philosophy, one who engaged with the principal ideas of his most important predecessor, René Descartes, and of his most influential critic, G. W. Leibniz. Unlike Descartes and Leibniz, Newton was systematic and philosophical without presenting a philosophical system, but, over the course of his life, he developed a novel picture of nature, our place within it, and its relation to the creator. This rich treatment of his philosophical ideas, the first in English for thirty years, will be of wide interest to historians of philosophy, science, and ideas.

ANDREW JANIAK is Assistant Professor in the Department of Philosophy, Duke University. He is editor of *Newton: Philosophical Writings* (2004).

NEWTON AS PHILOSOPHER

ANDREW JANIAK

Duke University

CAMBRIDGE
UNIVERSITY PRESS

CAMBRIDGE UNIVERSITY PRESS

Cambridge, New York, Melbourne, Madrid, Cape Town, Singapore, São Paulo,
Delhi, Dubai, Tokyo

Cambridge University Press
The Edinburgh Building, Cambridge CB2 8RU, UK

Published in the United States of America by Cambridge University Press, New York

www.cambridge.org
Information on this title: www.cambridge.org/9780521862868

First published 2008
Third printing 2009

Printed in the United Kingdom at the University Press, Cambridge

A catalogue record for this publication is available from the British Library

Library of Congress Cataloguing in Publication data
Janiak, Andrew.
Newton as philosopher / by Andrew Janiak.
p.   cm.
Includes bibliographical references (p. 179).
ISBN 978-0-521-86286-8
1. Newton, Isaac, Sir, 1642–1727 – Philosophy.   2. Philosophy of nature – History – 17th century.
3. Philosophy of nature – History – 18th century.   I. Title.
B1299.N34J36   2008
192–dc22   2007049998

ISBN 978-0-521-86286-8 hardback

Contents

Preface

This is a work in the history of philosophy. But it analyzes a figure considered first and foremost a mathematician and physicist. In recent years scholars have emphasized the importance of the complex interactions between philosophy and the natural sciences in the modern period. This study adopts a parallel perspective, attempting to shed new light on Isaac Newton by situating him within a rich philosophical milieu. I argue that this approach reflects Newton's own self-conception. He considered Descartes to be his most important predecessor among the myriad natural philosophers of the seventeenth century, and his principal contemporary interlocutors include many prominent philosophical figures, for instance Robert Boyle, Samuel Clarke, Roger Cotes, and John Locke in England, and Christiaan Huygens and Gottfried Wilhelm Leibniz on the Continent. For his supporters and detractors alike Newton single-handedly placed crucial topics on the philosophical agenda of the eighteenth century. It is hoped that this book will illuminate the philosophical aspects of Newton's work, and also bring a fresh perspective to key themes in the development of early modern philosophy.

I did not set out to write a book on Newton. Originally I was working on Kant's theoretical philosophy with Michael Friedman. When I first came to his office about a decade ago, Michael suggested that we read the Leibniz–Clarke correspondence together in preparation for my work on Kant's conception of space. This immediately sparked my interest in all things Newtonian. I eventually wrote a dissertation on what I called Kant's Newtonianism under Michael's direction – its chapter on Newton was my first attempt to understand the issues discussed in this book. Perhaps more than anything else, Michael has shaped my conception of what philosophy is – he taught me that any philosophical progress is tentative, and that our understanding of a text can always be deeper.

My early work was greatly advanced by a history of science seminar on Newton taught by Nico Bertoloni Meli: with his significant help, my paper in his seminar eventually became my first article, on Newton's atomism and conception of divisibility. As a graduate student at Indiana I also had the pleasure of working closely with Fred Beiser and Paul Franks. It was Fred who encouraged me to study German more intensely in Berlin, and his astonishing range of historical and philosophical knowledge served as an inspiration to me. I first met Paul Franks while I was a graduate student at Michigan – his philosophical depth and originality immediately sparked my interest in Kant, and I was blessed to continue working with him after we both moved to Indiana. I owe my original interest in Kant not only to Paul, but also to the incomparable David Hills, whose knowledge will never be plumbed. Both Fred and Paul have publicly acknowledged the extraordinary constellation of faculty and graduate students working on Kantian topics in Bloomington in the late 1990s – I benefited greatly from participating in this group. The Kant reading group in Bloomington, run by Graciela De Pierris, was very important for my work. Among many other interlocutors I would especially like to thank Hess Chung and Christian Johnson, the two most insightful and knowledgeable graduate students I have ever known – I learned immensely from our almost nightly discussions over mediocre Chinese food and endless lukewarm tea at the famous Dragon restaurant. David Finkelstein, Karen Hanson, and Hindy Najman were also sources of philosophical wisdom in Bloomington. While I held a graduate fellowship at Tel Aviv University I learned much about Kant from Susan Neiman, who was a generous host during my year in Israel. I gave my very first public lecture at Tel Aviv, on Kant and Newton (of course). Once my graduate work was over, I was fortunate to receive a postdoctoral fellowship at the Dibner Institute at MIT while George Smith was its acting director. No one who knows George will be surprised to learn that he often took two or three hours out of his busy schedule to discuss Newton with me. There is no better interlocutor anywhere for discussing Newton. While at the Dibner I learned a great deal from the history of mechanics reading group, which included Moti Feingold, Al Martinez, Jim Voelkel, George, and Nico.

My colleagues at Duke have been extremely supportive – because of their open-mindedness, I was able to bracket my research on Kant and write this book. Among my many helpful Duke colleagues I would especially like to thank Robert Brandon, Fred Dretske, Tim Lenoir, Alex Rosenberg, David Sanford, Laurie Shannon, Barbara Herrnstein Smith, and Susan Sterrett for their comments on my work. For his official mentorship, and constant fruitful advice, I thank Tad Schmaltz; whenever I have a perplexing question, I simply walk next door.

Various aspects of this book have been presented at the following venues: the joint meeting of the History of Science Society and the Philosophy of Science Association in Vancouver; the fifth and sixth international congresses of the History of Philosophy of Science Group (HOPOS) in San Francisco and in Paris; the Cartesian circle at Irvine; early modern philosophy workshops at the University of Pennsylvania and Mansfield College, Oxford; Newton conferences at Leiden University and at the Van Leer Institute in Jerusalem; and the philosophy departments at Toronto and Tufts. Numerous colleagues and friends have given me helpful criticism and feedback over the years – in addition to those mentioned above, I would like to thank Donald Ainslie, Lanier Anderson, Liz Anderson, Jeffrey Barrett, Zvi Biener, Katherine Brading, Justin Broackes, Janet Broughton, Anjan Chakravartty, Graciela De Pierris, Karen Detlefsen, Rob DiSalle, Mary Domski, Lisa Downing, Katherine Dunlop, Mordechai Feingold, Alan Gabbey, Daniel Garber, Don Garrett, Niccolo Guicciardini, Bill Harper, Sally Haslanger, Gary Hatfield, Barbara Herman, Nick Huggett, Rob Iliffe, Andrew Jainchill, Dana Jalobeanu, Lynn Joy, Sukjae Lee, Martin Lin, Paul Lodge, Jeff McDonough, Ernan McMullin, Scott Mandelbrote, Mohan Matten, Sy Mauskopf, Christia Mercer, Alan Nelson, Bill Newman, Eric Schliesser, Alison Simmons, Ed Slowik, Chris Smeenk, Kyle Stanford, Howard Stein, Friedrich Steinle, Daniel Sutherland, Jackie Taylor, Karin Verelst, Daniel Warren, Jessica Wilson, and John Young.

I owe a special debt of gratitude to the following people, who read the entire draft of the manuscript: Karen Detlefsen, Paul Franks, Michael Friedman, Daniel Garber, Sean Greenberg, Eric Schliesser, Tad Schmaltz, and George Smith. Their comments have proven tremendously helpful. Michael and George provided especially insightful and detailed suggestions.

When I published a collection of Newton's philosophical writings for Cambridge University Press, Hilary Gaskin was a wonderful editor; she has been once again with this volume. I am very grateful for her support of my work. I would also like to thank two anonymous referees who read the manuscript for Cambridge and made many fruitful suggestions. Many thanks to my research assistants, Shame Chikoro and James Abordo Ong.

The research for this book was supported financially by the following sources: the School of Historical Studies at Tel Aviv University; the Dibner Institute for the History of Science and Technology at MIT; the Josiah Charles Trent Memorial Foundation; and, at Duke, the Andrew W. Mellon Faculty Fund, the Franklin Humanities Institute, the Center for European Studies, the Center for Medieval and Renaissance Studies, the Patterson endowment in the Department of Philosophy, the Vice-Provost for

International Affairs, and the Vice-Provost for Interdisciplinary Studies. A grant from Duke's Arts and Sciences Committee on Faculty Research enabled me to use the Dibner Institute's world-class Grace Babson collection of Newton materials on several occasions, and the Institute's staff, especially Anne Battis, were very helpful. I finished this book while on a junior faculty research leave – thanks to Deans George McLendon and Gregson Davis for that crucial support, and to the incomparable staff of Duke's Franklin Humanities Institute for their hospitality during my leave.

Lastly and most importantly: I dedicate this book to Rebecca Luna Stein, my partner, and to Isaac Janiak Stein, our son. Rebecca's intellectual courage inspires me; she is a constant source of insight and also of love. Without her I would have lacked the courage to write this book. True to his name, Isaac brings great laughter to our lives each day. Raising him with Rebecca has been the most philosophically potent experience of my life.

Note on texts and translations

All translations are my own unless otherwise noted. In texts written in English in the seventeenth or eighteenth centuries I have expanded abbreviations and modernized spelling and capitalization (except when the meaning might be altered). All emphasis in quotations is in the original unless otherwise indicated

In citing texts I use the following abbreviations throughout:

Correspondence – *The Correspondence of Isaac Newton*, ed. H. W. Turnbull *et al.*, Cambridge: Cambridge University Press, 1959–77.

De Gravitatione – refers to the new translation by Christian Johnson in *Philosophical Writings* (see below).

Die philosophischen Schriften – *Die philosophischen Schriften von G. W. Leibniz*, ed. C. Gerhardt, Berlin: Weidmann, 1890.

Leibniz-Clarke Correspondence – Many editions are available; cited throughout by reference to letter and section numbers. For example, the tenth section of Leibniz's third letter is cited as "L 3:10," and the eighth section of Clarke's fifth letter as "C 5:8." For Leibniz's French, I use the text in *Die philosophischen Schriften* (see above).

Opticks – Isaac Newton, *Opticks, Or A Treatise of the Reflections, Refractions, Inflections & Colours of Light*, based on the fourth edition of 1730, New York: Dover, 1952.

Philosophical Writings – Isaac Newton, *Philosophical Writings*, ed. Andrew Janiak, Cambridge: Cambridge University Press, 2004.

Principes mathématiques – Isaac Newton, *Principes mathématiques de la philosophie naturelle*, trans. Madame la Marquise du Chatelet, Paris: Desaint & Saillant, 1749.

Principia – Isaac Newton, *The Principia: Mathematical Principles of Natural Philosophy*, trans. and ed. I. Bernard Cohen and Anne Whitman, with the assistance of Julia Budenz, Berkeley: University of California Press, 1999.

Principia Mathematica – Isaac Newton, *Isaac Newton's Philosophiae Naturalis Principia Mathematica*, the third edition (1726) with variant readings, ed. Alexandre Koyré and I. Bernard Cohen, with the assistance of Anne Whitman, Cambridge, MA: Harvard University Press, 1972.

Principia Philosophiae – René Descartes, *Principia Philosophiae*, in *Œuvres de Descartes*, ed. Charles Adam and Paul Tannery, vol. VIII-1, Paris: Vrin, 1982.

Newton as philosopher, the very idea

What does it mean to treat Newton as a philosopher?[1] We cannot identify any overarching philosophical position with Newton, as we identify dualism with Descartes, monism with Spinoza, or even classical empiricism with Locke. Newton never wrote a systematic philosophical treatise of the order of the *Meditations*, the *Ethics*, or the *Essay*. This was a significant choice, for he was perfectly familiar with such treatises, having already analyzed some in his youth. Even in his voluminous correspondence, and in his most important unpublished manuscripts, we do not find a systematic engagement with metaphysical issues akin to Descartes's. Newton positioned himself as a strong critic of Cartesianism, but his response to Descartes is significant as much for its lacunae as for its central claims. Thus these canonical philosophical figures, with their canonical texts, cannot serve as a model here.

To treat Newton as a philosopher might simply be to avoid an anachronistic characterization of his intellectual milieu. As scholars of the early modern period regularly note, the intellectual categories and disciplines of Newton's day – which ranges, roughly, from 1660 until 1730 – differ radically from our own. What we would consider to be separate fields of study – for instance, aspects of what we categorize as philosophy, especially metaphysics and epistemology, the physical sciences, and even theology – were interwoven into one overarching field called natural philosophy.[2]

---

[1] In a recent essay Calvin Normore raises this issue in passing, writing: "One of the most annoying philosophical questions is 'What is Philosophy?' Of course Philosophy is what Philosophers (*qua Philosophers*) do but who are the Philosophers? There is always Socrates but what about Isocrates? Leibniz, clearly, but what of Newton?" – Normore, "What is to be Done in the History of Philosophy," 75. See also the helpful discussion in Gabbey, "Newton, Active Powers and the Mechanical Philosophy," 329–31; Gabbey and I essentially agree on the best way to characterize Newton's status as a "philosopher."

[2] To my knowledge Edward Grant has written the only history of natural philosophy (see n. 4 below). On the very idea of natural philosophy, see Stein, "On Philosophy and Natural Philosophy in the Seventeenth Century," section 2 of Hatfield, "Was the Scientific Revolution a Revolution in Science?", Shapiro, *Fits, Passions and Paroxysms*, 26–40, and the incomparable treatment in Funkenstein, *Theology and the Scientific Imagination*. For an influential pre-Newtonian conception,

Hence to treat Newton as a philosopher in a historically accurate way might be to treat him as a natural philosopher, rather than more narrowly as a scientist, physicist, or mathematician.[3] Since Newton's *magnum opus* is called *The Mathematical Principles of Natural Philosophy*, there is little doubt that this attitude reflects his own self-conception.[4]

Yet this is only the beginning of the story. Newton certainly conceived of himself as a natural philosopher, among other things, but a brief glance at his *Principia* – including its full title – reveals the fundamental importance

see the Cartesian treatise by Rohault, *System of natural philosophy*, vol. I: 20–1. Rohault is especially concerned to indicate that natural philosophy analyzes objects within their natural state, where the contrast class is not the quasi-Aristotelian "violent" state of a thing, but rather some state that God creates against the ordinary course of nature. Hence Rohault explicitly brackets miracles, and what he calls "the mysteries of faith."

[3] As is well known, the word "scientist" was coined in 1833 by William Whewell when reviewing Mary Somerville's work (see Danielson, "Scientist's Birthright"); thus Somerville was likely the first person ever to be called a scientist. However, the term "science" was often used in the early modern period – see n. 4.

[4] In *A History of Natural Philosophy* Grant argues that we should not be "misled" by Newton's title, for he "might have used either of two medieval and early modern synonyms for natural philosophy, namely, 'natural science' (*scientia naturalis*) or 'physics' (*physica*)," to produce, respectively, the titles "Mathematical Principles of Natural Science" and "Mathematical Principles of Physics" (314). Grant contends that Newton kept his famous title – rather than the title *De Motu Corporum libri duo* – because it would help Halley, who saw the *Principia* through the press, to sell more books. On 20 June 1686, Newton wrote to Halley as follows:

I designed the whole to consist of three books . . . The third I now design to suppress. Philosophy is such an impertinently litigious lady that a man had as good be engaged in law suits as have to do with her. I found it so formerly [he presumably means the 1670s optics disputes] & now I no sooner come near her again but she gives me warning. The two first books without the third will not so well bear the title of *Philosophiae naturalis Principia Mathematica* & therefore I had altered it to this *De Motu corporum libri duo*: but upon second thoughts I retain the former title. T'will help the sale of the book which I ought not to diminish now tis yours. (*Correspondence*, vol. II: 437)

This was written in response to Halley, who had written to Newton two weeks earlier that he ought to include book III because "the application of this mathematical part, to the system of the world; is what will render it acceptable to all naturalists, as well as mathematicians, and much advance the sale of the book" (*Correspondence*, vol. II: 434). For his part, Grant concludes: "the argument that Newton did not believe he was doing natural philosophy in the *Principia* gains credibility from Newton himself" (*A History of Natural Philosophy*, 315). But this is misleading. Newton ultimately included the third book of the *Principia*, and his substantive point in his letter – which coheres with Halley's substantive point in his letter – is plain: without the third book, in which Newton discusses what he calls "the system of the world," the first two books of the *Principia* could not very accurately be called natural philosophy; rather, they would be better described as two books on the motion of bodies. For Newton as for Halley, natural philosophy was not principally concerned with just any motion of bodies that was tractable through mathematical analysis; rather, it was concerned with the mathematical analysis of the motions of the bodies within our solar system – within nature as we perceive it in our vicinity of the universe. And that, of course, is part of a long tradition in natural philosophy. Newton's innovation is to provide a mathematical analysis of the motions of these bodies. On Halley's role in printing the *Principia* see Cohen, *Introduction to Newton's "Principia,"* 130–42; for Halley's relation to Newton, see the materials reprinted and discussed in Cohen and Schofield, *Isaac Newton's Papers and Letters*, 397–424, which includes Halley's review of the first edition of the *Principia* in the *Philosophical Transactions of the Royal Society*.

of the fact that his principles are "mathematical."[5] Newton titled his work to establish a replacement for Descartes's *Principles of Philosophy*, first published in Amsterdam in 1644, a text that Newton read carefully and kept in his personal library.[6] The differences between the two works are stark: whereas Descartes's text is familiar to historians of modern philosophy, with its focus on broadly conceived epistemic and metaphysical issues, Newton's text is a highly technical mathematical work that apparently ignores such issues altogether. Whereas Descartes's *Principia* attempts to account for an immense range of phenomena – tackling everything from global skepticism about human knowledge to God's immutability to the nature of heat, light, weight, and so on – Newton's text focuses specifically on the mathematical analysis of motion and the forces that cause it. The audiences of the two works differed accordingly: Descartes was comprehensible to anyone with a decent education in the codified Aristotelian corpus, or late Scholastic natural philosophy. In contrast, Newton's *Principia* was comprehensible only to the most sophisticated mathematicians. Descartes thought that metaphysics and physics could follow the same humanistic

---

[5] This is, of course, signaled in the very title of the *Principia*, but it characterizes his pre-*Principia* work as well. Newton had insisted on the importance of mathematics within natural philosophy already in his early optical research. Consider, for instance, this remarkable passage from his *Lectiones Opticae* of 1670, delivered as his inaugural Lucasian Professor lectures:

> the generation of colors includes so much geometry, and the understanding of colors is supported by so much evidence, that for their sake I can thus attempt to extend the bounds of mathematics somewhat, just as astronomy, geography, navigation, optics, and mechanics are truly considered mathematical sciences even if they deal with physical things: the heavens, earth, seas, light, and local motion. Thus although colors may belong to physics [*ad Physicam pertineant*], the science of them must nevertheless be considered mathematical, insofar as they are treated by mathematical reasoning. Indeed, since an exact science [*accurata scientia*] of them seems to be one of the most difficult that philosophy is in need of, I hope to show – as it were, by my example – how valuable mathematics is in natural philosophy. I therefore urge geometers to investigate nature more rigorously, and those devoted to natural science to learn geometry first. (*Optical Papers*, vol. I: 86–7; Shapiro trans.)

Isaac Barrow preceded Newton as the Lucasian Professor at Cambridge; for a helpful discussion of his influence on Newton's conception of the role of mathematics within natural philosophy, see Shapiro, *Fits, Passions and Paroxysms*, 30–40. For a brief discussion of the teaching of mathematics to Cambridge undergraduates in Barrow's and Newton's day, see Costello, *The Scholastic Curriculum at Early Seventeenth-Century Cambridge*, 102–4.

[6] When Newton wrote to Hooke in a famous letter, "If I have seen further it is by standing on the shoulders of giants," he included Descartes among them (*Correspondence*, vol. I: 416). Remarkably, Newton sometimes referred to the *Principia* as his "*Principia Philosophiae*" – for instance, in his (anonymous) "An Account," 180 and 198; cf. the notes of Cohen and Whitman as editors at *Principia*, 11. For discussion see especially the extensive chapter on Descartes and Newton in Koyré, *Newtonian Studies*, and also Cohen, "Newton and Descartes"; cf. Stein, "Newtonian Space–Time," and "Newton's Metaphysics." In *Elements of Early Modern Physics*, Heilbron writes that as of the mid-seventeenth century "Descartes then replaced Aristotle as the foil against which British physics tested its metal" (30).

methods, but for Newton physics was essentially mathematical.[7] Although both works belong to the seventeenth-century canon in natural philosophy, then, they represent two fundamentally distinct traditions.[8]

Newton eschews many of the issues that Cartesians placed at the center of natural philosophy, but not all of them. There is a danger of his overwhelming influence on physics in the eighteenth century blinding us to his own conception of how a mathematical investigation of natural phenomena might intersect with broader metaphysical concerns, such as God's relation to the physical world. Of course, one of the primary aspects of the *Principia*'s intellectual impact in the eighteenth century was the separation it effected between technical physics on the one hand, and philosophy on the other. In the hands of figures like Laplace and Lagrange, Newton's work led to the progressive development of Newtonian mechanics, and its practitioners embraced a conception of their discipline in which philosophical matters played little role. Yet these facts obscure Newton's own conception of his work. As he said repeatedly throughout his career, investigating the first cause is a proper part of natural philosophy.[9] Thus

---

[7] Newton and Descartes may also have differed in their understanding of mathematics, especially regarding geometry and geometric construction – see Domski, "The Constructible and the Intelligible." In this paper Domski helpfully indicates why Newton's famous discussion of geometry and mechanics in the author's preface to the first edition of his text (*Principia*, 381–2) may be misleading (1120–3).

[8] The treatment that the two works receive within modern scholarship reflects this fact: whereas Descartes's work is read almost exclusively by historians of philosophy, Newton's work is read almost exclusively by historians of science. This parallel masks a further point, however, for Newton's influence on modern mathematical physics is also reflected in part by the fact that historians of science, but not physicists, approach his work as a canonical text. For it is partly in virtue of Newton's success in founding the modern discipline of physics that contemporary physicists evince little interest in canonical texts, or in historical reconstructions. Newton himself already prefigured this perspective in the *Principia*, when he incorrectly attributed the first two laws of motion to Galileo, who certainly never articulated them in their Newtonian form (*Principia*, 424). But textual accuracy, and historical understanding, is presumably not Newton's aim in this passage; rather, Newton sees himself working within the same tradition as Galileo, and he attempts to further Galileo's work, for instance by providing a new understanding of why all bodies in free fall exhibit the same acceleration. On Galileo, see Bertoloni Meli, *Thinking with Objects*, 145–6.

[9] This represents Newton's consistent view, expressed in both published and unpublished materials, throughout his mature intellectual life. Among his published works see the General Scholium, which Newton added to the end of his text (*Principia*, 943) and the queries (*Opticks*, 369, 405). Among his unpublished works, see his first letter to Richard Bentley in 1692 (*Philosophical Writings*, 95–6), and his draft of the first edition preface to the *Opticks*, in which he writes: "One principle in Philosophy is the being of a God or Spirit infinite eternal omniscient omnipotent, & the best argument for such a being is the frame of nature & chiefly the contrivance of the bodies of living creatures" (McGuire, "Newton's 'Principles of Philosophy,'"183). Newton then proceeds to recount a design argument, one that appears elsewhere in his published works (e.g. in the General Scholium). The very next principle is the impenetrability of matter, followed by a third principle, which is expressed by the law of universal gravitation.

the key to treating Newton as a philosopher lies in discovering how the mathematical treatment of force and motion forms part of the same enterprise as an investigation of seemingly separate metaphysical issues, such as God's relation to the world.[10]

Yet even this characterization remains too narrow: Newton tackled numerous "metaphysical" topics separate from an analysis of the divine being. He did so, perhaps, as a matter of necessity. The astonishing success of Newtonian mechanics in the eighteenth century – a fact confronted by most of the major philosophers of that period – should not mislead us into thinking that Newton's work was immediately accepted by those in a position to assess it. On the contrary, the theory of gravity in the *Principia*, which first appeared in 1687, was not merely challenged on narrow mathematical or empirical grounds, but fundamentally rejected for its violation of the norms established by adherents of the mechanical philosophy, such as Leibniz and Huygens.[11] Thus in his correspondence after 1687, in the queries to the *Opticks*, and especially in the second edition of the *Principia* (1713), Newton was forced to defend his mathematical treatment of force and motion on fundamental metaphysical grounds.[12] These elements of Newton's work are crucial to the account in this book.

[10] For a significantly distinct perspective on these issues, see Cohen and Smith's introduction to their volume, *The Cambridge Companion to Newton*. See also the illuminating discussions in Hatfield, "Metaphysics and the New Science," 144–6, and in Friedman, "Metaphysical Foundations of Natural Science," 239.

[11] For his part Leibniz conceived of Newton as a metaphysician, although he contended that he had "little success in metaphysics" in a letter to J. Bernoulli in March of 1715 (*Correspondence*, vol. VI: 212–13). He responded to the *Principia* in his *Tentamen* of 1689 (discussed in ch. 2), in a significant 1693 letter to Newton (*Philosophical Writings*, 106–7, also discussed below), and in much of his correspondence late in life, especially the famous series of exchanges with Clarke in 1715 and 1716. Leibniz's criticisms of Newton are well known, but Huygens's may not be. He wrote to Leibniz in November of 1690 as follows: "Concerning the cause of the tides given by M. Newton, I am by no means satisfied, nor by all the other theories that he builds upon his principle of attraction, which seems to me absurd, as I have already mentioned in the addition to the *Discourse on Gravity*" (*Œuvres complètes*, vol. IX: 538). I first learned of this important letter from George Smith. Huygens maintained his adherence to the mechanical philosophy in the face of Newton's deviation from its norms in his theory of gravity – see especially his *Discours sur la cause de la Pesanteur* in *Œuvres complètes*, vol. XXI: 129–30, 159–60 (this is the original pagination, given marginally in the text). Newton was present when Huygens introduced his *Discours* to the Royal Society in 1689 (Westfall, *Never at Rest*, 488). Huygens and Leibniz each received a copy of the *Principia* from Newton himself – see *Never at Rest*, 472. It should be noted that Huygens also uncovered empirical reasons for doubting Newton's conclusion in book III of the *Principia* that gravity acts universally on all bodies: see the helpful account in Schliesser and Smith, "Huygens's 1688 Report," 5–7.

[12] Ironically, Newton's early optical work, published in the Royal Society's *Philosophical Transactions* in the 1670s, was similarly greeted with fundamental objections and, from Newton's point of view at least, with fundamental misinterpretations. This episode also spawned a series of responses from Newton regarding, among other things, the proper methods in optical research and theorizing.

The extensive debate about Newton's methods – which spawned the famous correspondence between Leibniz and Clarke in the early eighteenth century – suggests that there were two revolutionary developments in seventeenth-century natural philosophy. One involved a shift from neo-Aristotelian or "Scholastic" natural philosophy to Cartesianism, and the other a shift from mechanistic to Newtonian natural philosophy. This second shift raised philosophical problems as profound as those accompanying the first. The first revolutionary development involved significant conceptual continuity. Although the Aristotelians in the late sixteenth and early seventeenth centuries resisted the idea that everything in nature could be explained mechanically – arguing that some phenomena must be explained through a combination of matter and form – they would certainly have found that idea intelligible. The neo-Aristotelians had a mechanics that, although not foundational like Cartesian mechanics, and although not part of an overarching "mechanical philosophy," did involve the explanation of some processes in terms of matter and motion, where matter is characterized solely by the size and shape of its parts. Descartes then made the widespread intelligibility of his mechanical explanations a hallmark of his work.[13]

With the revolutionary shift from mechanistic to Newtonian natural philosophy after 1687, however, we find a fundamental lack of conceptual continuity. Just as the mechanists pronounced Scholastic explanations employing matter and form to be unintelligible, they treated Newton's explanation of various gravitational phenomena as equally incoherent. The criticisms mirrored one another in a significant way: Leibniz maintained for years that Newton had revived precisely the incoherent Scholastic talk of occult qualities that the mechanists had attempted to expunge permanently

---

Although I discuss some of these issues briefly in ch. 2, largely to underscore the parallels between the debates in optics and those in physics, this book focuses especially on the latter. For further details, see Harper and Smith, "Newton's New Way of Inquiry."

[13] See especially the discussion in Garber, "Descartes, Mechanics, and the Mechanical Philosophy." When reflecting on the character of his own work, Descartes writes in section 200 of *Principia Philosophiae*: "I have used no principles in this treatise which are not accepted by everyone; this philosophy is not new but is very old and very common." He then elaborates:

I have considered the shapes, motions and sizes of bodies and examined the necessary results of their mutual interaction in accordance with the laws of mechanics, which are confirmed by reliable everyday experience. And who has ever doubted that bodies move, have various sizes and shapes, and various different motions corresponding to these differences in size and shape; or who doubts that when bodies collide bigger bodies are divided into many smaller ones and change their shapes? (*Principia Philosophiae*, vol. VIII-1: 323)

from natural philosophy.[14] Newton always embraced the familiar view that God represents a fundamental component of our understanding of nature, expressing that view even in the *Principia*'s first edition.[15] But with the mechanistic charges of unintelligibility, he was pressed into more extensive debates concerning the metaphysical presuppositions and implications of his mathematical treatment of force and motion. In the course of these discussions Newton consistently articulated a compelling and overarching conception of the relation between mathematical physics on the one hand, and more clearly metaphysical concerns on the other. These lie at the center of this book.

Newton's response to the intellectual controversy generated by the *Principia*, in turn, helps to illuminate the narrowness of his metaphysical ruminations. Newton lacks a metaphysical system that addresses the major topics raised by his most important predecessor, Descartes, or his most insistent critic, Leibniz. His discussions are largely limited to questions about the ontology of space and time, the laws of motion and the forces that cause motion, our knowledge of matter within physics, and God's relation to the physical world.[16] This reflects Newton's famous reluctance to engage

---

[14] One of Leibniz's most colorful discussions occurs in a letter he wrote to Nicolas Hartsoeker, later published in English translation in the *Memoirs of Literature* in 1712:

Thus the ancients and the moderns, who own that gravity is an *occult quality*, are in the right, if they mean by it that there is a certain mechanism unknown to them, whereby all bodies tend towards the center of the earth. But if they mean that the thing is performed without any mechanism by a simple *primitive quality*, or by a law of God, who produces that effect without using any intelligible means, it is an unreasonable occult quality, and so very occult, that it is impossible that it should ever be clear, though an angel, or God himself, should undertake to explain it. (*Philosophical Writings*, 112)

He makes a similar charge in his last letter to Clarke (L 5: 113, *Die philosophischen Schriften*, vol. VII: 417). I discuss this issue in depth in ch. 4.

[15] In corollary five to proposition 8 of book III, Newton discusses the density of the planets and their placement relative to the sun, arguing: "God therefore placed the planets at different distances from the Sun so that according to their degrees of density they may enjoy more or less of the Sun's heat" – see *Principia Mathematica*, vol. II: 583 note. The reference to God was removed in the second edition (*Principia Mathematica*, vol. II: 582–3; *Principia*, 814), but that edition included the new General Scholium, which contains the following remark: "This most elegant system of the sun, planets, and comets could not have arisen without the design and dominion of an intelligent and powerful being" (*Principia*, 940). Although Newton removed the original passage from the second edition, at least one of his confidants, Samuel Clarke, continued to cite it years later – see his Boyle Lectures for 1704, *A Demonstration of the being and attributes of God*, 82. For an excellent discussion of these issues, see Cohen, "Isaac Newton's *Principia*, the Scriptures, and the Divine Providence," 529ff.

[16] Intriguingly, *De Gravitatione* – an unpublished manuscript in which Newton presents numerous criticisms of Cartesian natural philosophy – is not an exception to this point, for although it was most probably written before 1687, it concerns the somewhat narrow range of issues that Newton tackles in later published texts and correspondence. Thus in *De Gravitatione* Newton discusses motion, the ontology of space and time, the nature of material bodies, mind–body dualism, and God, rejecting

in intellectual disputation – evident already in his earliest optical writings in
the 1670s – and also his attempt to rebut mechanist criticisms, for Newton
and the mechanists were engaged in a philosophical dispute regarding
physics itself. Nonetheless, as we shall see, Newton inherits a series of
fundamental concepts from the seventeenth-century metaphysical tradition
represented by figures such as Descartes. He consistently articulates his
most basic views by employing the concepts God, substance, and action.[17]
Hence he does not typically employ novel concepts in his work within
natural philosophy; the novelty of his views lies in part in his transformation
of these concepts in sometimes controversial ways.

Despite his use of well-worn metaphysical concepts – such as God and
substance – that are central to the Cartesian system, Newton's basic orientation toward philosophical issues differs fundamentally from that evinced by
Descartes. Given Descartes's view of the relation between metaphysics
and physics, responding to skepticism is as significant for the latter as it is
for the former. Metaphysics serves as the foundation for physics – *Principia
Philosophiae* presupposes basic metaphysical principles concerning (e.g.)
the essence of body and the nature of God. These metaphysical principles,
in turn, are firmly established by surviving a confrontation with a fundamental skepticism about our knowledge of nature; the defeat of this skepticism signals that we can be certain of our basic metaphysical knowledge.
Since physics presupposes the first principles of metaphysics, and since
metaphysics is characterized by a struggle to establish its own authority,
our knowledge in physics hangs in the balance when we confront radical
skepticism in the *Meditations*. For Newton, in contrast, global skepticism is
irrelevant – he takes the possibility of our knowledge of nature for granted.[18]
Instead, the primary epistemic questions confronting us are raised by

---

various Cartesian views along the way, but he ignores other issues in Cartesian metaphysics, such as
the proofs for the existence of God, the proper response to skepticism, the question of innate ideas,
etc. For a helpful suggestion about the dating of *De Gravitatione* see Feingold, *The Newtonian
Moment*, 25–6. For an extensive discussion of *De Gravitatione*, see Steinle, *Newtons Entwurf "Über die
Gravitation."*

[17] Many thanks to Michael Friedman for discussion of this point.

[18] So if science is separated from philosophy in Newton's hands, it is not through the defeat of
skepticism, but through its rejection as a fundamental problem confronting our knowledge of nature.
This may be partially definitive of the modern distinction between science and philosophy. Whereas
modern science continued to ignore skepticism, taking it for granted that we can achieve knowledge
of nature, philosophy continued to be preoccupied by skepticism of various varieties, assuming that
we must confront it in one way or another. Cf. Bloch, *La Philosophie de Newton*, 492. Thanks to Sean
Greenberg for discussion of this point.

physical theory itself. Thus Newton eschews one of the animating features of the early modern philosophical debate.[19]

So we might say that Newton is systematic and philosophical without articulating a philosophical system.[20] He never embraces a global metaphysical position such as dualism or monism, and he never presents an overarching theory of knowledge or response to global skepticism. But he deals systematically with those elements of metaphysics that are intimately connected with his work in mathematical physics. In the course of defending his work from the mechanists he presents a novel conception of the relation between physics and metaphysics. This book aims to clarify that conception.

The rest of the book proceeds as follows. In ch. 2 I focus on three interpretations of Newton's conception of the relation between physics and metaphysics, each of which responds explicitly to the issues raised above. The chapter's dialectic can be expressed in this way: whereas the first interpretation, which originates in the eighteenth century, conceives of Newton as a kind of anti-metaphysician, the second attempts to account for his anti-metaphysical remarks by arguing that he transforms traditional metaphysical questions into empirical ones. The third interpretation, which I defend throughout this book, acknowledges the significance of Newton's transformation of certain metaphysical questions into empirical ones, but also discusses his commitment to an overarching metaphysical picture of God's relation to the physical world.

The following three chapters emphasize Newton's novel approach to traditional metaphysics, while highlighting his commitment to a metaphysical framework centering on God. That framework is minimalist, but involves a clear conception of God's relation to the world, the nature of action, and the ontology of space and time. In ch. 3 I argue that Newton can retain his traditionalist rejection of action at a distance while postulating that all material objects in the world interact gravitationally in accordance with the theory in the *Principia*. Thus Newton is able simultaneously to reject the

---

[19] This difference is reflected in the Cartesian and Newtonian texts themselves. Descartes makes allowances for those who are not initiates in his project, beginning his *Meditations* from a universally accessible standpoint – the perspective of a knowing agent with ordinary sensory beliefs that can be called into question through universally accessible procedures, such as the contemplation of the possibility that one believes one is awake, but is actually dreaming. Newton, in contrast, makes no allowances for the lay reader, beginning his deductions in the *Principia* from a standpoint that is accessible only to mathematicians. This is also reflected in the point made in n. 18.

[20] Maclaurin describes this aspect of Newton's achievement in a typically favorable – and generally accurate – way at the end of book I of *An Account*, 95–6. See also Gabbey, "Newton, Active Powers, and the Mechanical Philosophy," 335. Thanks to Eric Schliesser for discussion of this point.

mechanical philosophy, while avoiding a kind of physical action that he famously calls "inconceivable." I then argue in ch. 4 that Newton's rejection of mechanist principles runs even deeper – it is reflected in his conception of the essence of matter, and in his understanding of the knowledge of matter expressed in physical laws. This indicates the strength of Newton's commitment to a physical approach to metaphysical problems, for he transforms the metaphysical discussion of physical action and the nature of matter among the mechanists into an empirical investigation. Despite these revolutionary views, however, Newton maintains his allegiance to a fundamental metaphysical framework centered on his (often provocative) conception of God. This framework is evident especially in Newton's famous discussion of absolute space, the focal point of ch. 5. Finally, ch. 6 employs my interpretation of Newton's metaphysical framework to illuminate his refusal to embrace action at a distance, despite significant empirical pressure originating with the *Principia* itself. Throughout his mature intellectual life, I argue, Newton articulated a minimalist but significant conception of God's action within the spatiotemporal world of the natural philosopher. The depth of his commitment to this conception is underscored by the fact that he never wavered from it, even in the face of the substantial controversies that it generated over the years. Newton was, in that regard, an impressively consistent thinker.

2

Physics and metaphysics: three interpretations

What is taught in metaphysics, if it is derived from divine revelation, is religion; if it is derived from phenomena through the five external senses, it pertains to physics; if it is derived from knowledge of the internal actions of our mind through the sense of reflection, it is only philosophy about the human mind, and its ideas as internal phenomena likewise pertain to physics. To dispute about the objects of ideas except insofar as they are phenomena is dreaming. In all philosophy we must begin from phenomena and admit no principles of things, no causes, no explanations, except those which are established through phenomena [In omni Philosophia incipere debemus a Phaenomenis, & nulla admittere rerum principia nullas causas nullas explicationes nisi quae per phaenomena stabiliuntur]. And although the whole of philosophy is not immediately evident, still it is better to add something to our knowledge day by day than to fill up men's minds in advance with the preconceptions of hypotheses.

Draft of preface to second edition (*Principia*, 54)

Numerous interpretations of Newton's work, and of his most general philosophical views, have appeared over the past two centuries. Indeed, by the middle of the eighteenth century, only two decades after Newton's death in 1727, there had already been a huge rash of English and French commentaries published in England and on the Continent.[1] To establish a focus, this chapter considers three particularly salient interpretations of Newton's metaphysical views that explicitly address the issues raised in ch. 1 – that is, each interpretation reacts explicitly to the fact that Newton presented his work as a mathematical alternative to the non-mathematical natural philosophy of Descartes's *Principia Philosophiae*. The dialectic that pushes from the first interpretation to the second, and eventually to my own

---

[1] For instance, s'Gravesande, *Mathematical Elements of Natural Philosophy*, Maclaurin, *An Account of Sir Isaac Newton's Philosophical Discoveries*, Châtelet and Clairaut's commentary appended to the second volume of *Principes mathématiques*, and Voltaire, *Eléments de la philosophie de Newton*.

(third) view, ought to illuminate the principal questions that this book attempts to answer.

The first interpretation, which appeared in the early eighteenth century and which retains currency today, contends that Newton fundamentally eschewed the very metaphysical issues that animated Descartes's work, focusing instead on empirical and mathematical topics that could be solved using recognizably modern scientific techniques.[2] Hence the famous methodological slogan, *hypotheses non fingo*, emerges in this interpretation as expressing a kind of metaphysical agnosticism. Something akin to this interpretation may be the most natural reading of Newton: if one can analytically distinguish his metaphysical from his mathematical and scientific achievements, the latter certainly dwarf the former. Although Newton deals with some of the metaphysical topics on the Cartesian agenda, his views may represent a mere historical residue, one that plays no substantive role in his principal work in the *Principia*.

The second interpretation envisions a somewhat more complex relationship between Newton's work in metaphysics and in physics, one that can be fruitfully characterized by distinguishing it from the Cartesian conception of that relationship. It is not that Newton simply eschews the principal metaphysical issues raised by Descartes; rather, he transforms what had been considered *a priori* issues concerning (e.g.) causation into empirical issues that can be settled on the basis of physical theory itself.[3] Part of the motivation for this reading is to recast Newton's explicitly metaphysical discussions – for instance, in the Scholium on space and time and in the General Scholium in the *Principia* – as reflective of Newton's overarching conception of the logical priority of physics over metaphysics. Given that it can make good sense of Newton's explicit discussion of metaphysical issues in his central texts, this second reading might be thought to represent an advance over the first reading outlined above.

---

[2] The first interpretation is not intended to capture the views of any one reader of Newton's work; rather, aspects of it, and versions of it, were defended by various writers beginning already in the early eighteenth century and continuing well into the twentieth. For instance, some of Samuel Clarke's defenses of Newton in his famous correspondence with Leibniz in 1715–16 evidence aspects of this interpretation, as I discuss in ch. 3. Jammer, in *Concepts of Force*, and Cohen in *The Newtonian Revolution*, defend several aspects of it, as does Bloch in *La Philosophie de Newton* (cf. Metzger's discussion of this aspect of Bloch in *Attraction universelle*, part 1: 13–16). Other aspects of the reading can be found in Harman, *Metaphysics and Natural Philosophy*, 1ff.

[3] Unlike the first interpretation, I take this second reading of Newton to be represented in the recent work of two scholars, Howard Stein and, following Stein, Robert DiSalle. However, not every aspect of the interpretation, as it is outlined in this chapter, can be found in both of their writings – for instance, Stein seems to differ from DiSalle on the question of action at a distance in Newton's thought. I discuss this issue in more depth below.

Just as the first interpretation may have difficulty accounting for important passages in Newton's *Principia*, the second interpretation may have difficulty in accounting for an explicitly articulated aspect of Newton's conception of natural philosophy in that text. Newton famously proclaims in the second edition of the *Principia* that studying God is a proper part of natural philosophy (*Principia*, 943).[4] But if a conception of the divine being lies at the center of, or perhaps animates, Newton's work in natural philosophy, it may be difficult to support the second reading of his attitude toward the metaphysical tradition represented by figures such as Descartes. Thus the third interpretation, which is defended in this book, places Newton's conception of God at its center. On my reading Newton envisions a complex interplay between physics and metaphysics – he neither eschews the latter in favor of the former, nor reorients the latter in terms of the former. Newton's metaphysical worldview is bifurcated into two intimately related but distinct elements: the first element expresses his conception of God's active role within the natural world, forming a fundamental metaphysical framework that is immune to revision from physics; the second element, concerning the material world we inhabit, lies within this framework and fully reflects the results of physical theory. Thus, for instance, the metaphysical framework includes a characterization of God's fundamental relation to space and time – one not open to revision through empirical research – but remains agnostic on the proper conception of space's relation to ordinary material objects; that conception must reflect physical theory. In this way I attempt to capture the fundamental insight of the second interpretation of Newton's metaphysics while maintaining fidelity to Newton's own explicit characterizations of God's place within the world.

The three interpretations outlined in this chapter constitute distinct perspectives on Newton's relation to the metaphysical tradition of the seventeenth century, but they need not differ on the proper characterization of that tradition itself. Of course, there are various streams within metaphysics in the late seventeenth century, and the three readings of Newton can agree in taking only one of them to be salient for understanding him. That is, they need not differ on the interpretation of Newton's

---

[4] Although this aspect of Newton's thought – as it is expressed, for instance, in the General Scholium (1713) to the *Principia* – is well known, there is a debate concerning the actual place of the study of God within natural philosophy in the early modern period in general, and in Newton's system in particular. See Cunningham, "How the *Principia* Got its Name," "The Identity of Natural Philosophy," and Grant, "God and Natural Philosophy," which responds to Cunningham.

metaphysics solely because they differ on the question of what counts as a metaphysical view – the debate need not be merely semantic.

If we focus solely on the metaphysical conversations available to Newton himself, given his training at Trinity College in the early 1660s, his reading of the "moderns," and his correspondence and writings, we can separate out several distinct streams.[5] For those with at least a quasi-Aristotelian conception, metaphysics concerns an inquiry into being *qua* being, and not, for instance, an inquiry into natural beings. For others metaphysics concerns especially non-physical beings, such as God, angels, and the soul. And for some seventeenth-century natural philosophers, including some broadly "Cartesian" thinkers, metaphysics involves three principal matters: first, an investigation into the "first principles" that enable our knowledge of natural phenomena; second, an investigation into the basic components of the natural world; and third, an investigation of God's relationship to nature. They might focus on questions such as these. Are forces or powers genuine elements of the natural world, or is the world composed solely of bits of matter in motion? What types of causation underwrite change in the natural world? What is the nature of space and time, and what is God's relation to them?[6] This tradition is the salient one for all three interpretations outlined in this chapter. This is uncontroversial, for Newton is silent on many of the principal topics addressed by thinkers working within the other traditions. So the three interpretations differ on this question: does Newton answer the principal questions that animate what we might call the metaphysics of natural philosophy?[7]

---

[5] My discussion here reflects a basic agreement with Gabbey's historically precise account in "Newton, Active Powers, and the Mechanical Philosophy," 330–1. The three streams are perhaps only partially distinct, since they may overlap from the perspective of some philosophers in this period, and this list is not exhaustive (thanks to Karen Detlefsen for discussion of this point). On Newton's education, see especially the magisterial treatment in Westfall, *Never at Rest*, and see the extant version of Newton's undergraduate notebook from Trinity College, which indicates his broad familiarity with "moderns" such as Descartes, Boyle, and Hobbes (*Certain Philosophical Questions: Newton's Trinity Notebook*). Cf. also Costello, *The Scholastic Curriculum at Early Seventeenth-Century Cambridge*, 4.

[6] Stein provides a similar description in "Newton's Metaphysics," 256–7, emphasizing that Newton and his critics evidently do not distinguish between metaphysics and epistemology as a contemporary reader might.

[7] In Newton's day, of course, metaphysics was often taken to be distinct from natural philosophy (Gabbey, "Newton, Active Powers, and the Mechanical Philosophy," 329–30), just as many pre-Newtonians took natural philosophy to be distinct from mathematics, a circumstance that Newton himself bemoaned (e.g. in his Lucasian Lectures – *Optical Papers*, vol. I: 86–7). My aim here is simply to highlight what to contemporary eyes appear to be discussions of metaphysical issues within the context of developing an analysis of various natural phenomena.

There is a long and influential tradition of interpreting Newton as engaging in foundational debates within natural philosophy precisely to remove as many metaphysical elements as possible. That is to say, according to a traditional interpretation of Newton that hails from at least the early eighteenth century, we ought to recognize that his primary contribution to debates concerning the proper methods in natural philosophy, and to debates with the "mechanists" concerning the status of forces, was to have strenuously denied that natural philosophy ought to answer such metaphysical questions. On this view Newton did discuss the "first principles" of the study of nature in a way that he never discussed (e.g.) substance dualism; but the point in addressing the former was precisely to insulate natural philosophy from metaphysics and its concern with issues such as the latter. If one takes metaphysical questions concerning body, force, and motion to be inherent to natural philosophy during the period of Newton's *Principia*, then we can cast this view in somewhat different terms; in removing the metaphysical elements of natural philosophy, Newton helped to establish physics as a separate discipline. Or if you like Newton helped to transform natural philosophy into modern physics.[8]

This very general characterization of the anti-metaphysical reading of Newton's work, in turn, can be rendered more specific by considering a well-known aspect of the *Principia*'s methodology. Newton famously proclaims that book I of the *Principia* employs only a "mathematical treatment of force." He had introduced this method already in the definitions that open the *Principia* and that precede the introduction of the laws of motion. Consider, for instance, Newton's discussion of the three different "quantities" or measures – motive, accelerative, and absolute – of centripetal force in definition eight. Having noted that these quantities of forces may be called motive, accelerative, and absolute forces, Newton writes that absolute forces may be referred

to the center as having some cause without which the motive forces are not propagated through the surrounding regions, whether this cause is some central body (such as a lodestone in the center of a magnetic force or the earth in the center of a force that produces gravity) or whether it is some other cause which is not apparent. This concept is purely mathematical, for I am not now considering the physical causes and sites of forces. (*Principia*, 407)[9]

---

[8] See especially Cohen and Smith's introduction to *The Cambridge Companion to Newton*.
[9] For the original, see *Principia Mathematica*, vol. I: 45.

So the concept of an absolute (centripetal) force, it seems, is a mathematical one; this means, among other things, that this notion represents an abstraction away from the physical basis of the force, whether it be a magnet or some other material object.

Moreover, Newton famously generalizes this point at the end of his discussion of definition eight, writing:

> Further, it is in this same sense that I call attractions and impulses accelerative and motive. Moreover, I use interchangeably and indiscriminately words signifying attraction, impulse, or any sort of propensity toward a center, considering these forces not from a physical but only from a mathematical point of view. Therefore, let the reader beware of thinking that by words of this kind I am anywhere defining a species or more of action or a physical cause or reason, or that I am attributing forces in a true and physical sense to centers (which are mathematical points) if I happen to say that centers attract or that centers have forces. (*Principia*, 408)

So with respect not just to what he has called the "absolute quantity" of centripetal force, but with respect to centripetal forces generally, Newton appears to be contending that he considers forces only from a mathematical point of view.

But what precisely does it mean to treat forces purely from a mathematical point of view? According to the anti-metaphysical interpretation – one defended, in part, by Clarke and by Berkeley already in the early eighteenth century[10] – we should understand these passages in the *Principia* as expressing a stringent causal agnosticism. Newton writes that in using terms like "attraction" he does not intend to be defining a "species or mode of action or a physical cause or reason." For some of his interpreters, then, it is rather natural to think that from his mathematical point of view, we can (e.g.) treat the motions of various planetary bodies as if they arose from a centripetal force; but this is merely a mathematically precise way of presenting the phenomena, and not a causal claim concerning the motions themselves.

Given Newton's caveats and pronouncements at the very opening of the *Principia*, then, it seems that the anti-metaphysical reading of Newton's mathematical treatment of force is a reasonable one. Moreover, as Clarke and Berkeley were aware, it is also supported by a particularly difficult and well-known implication of the most prominent metaphysical reading of Newton's treatment of force. To sketch the problem with such a reading, consider this: if we disregard Newton's caveats following definition eight by contending that some types of centripetal force – gravity perhaps being most prominent among them – are in fact real forces from Newton's point

---

[10] I discuss Clarke's and Berkeley's views briefly in ch. 3.

of view, we immediately face the problem of action at a distance.[11] If, for instance, we take attraction not in Newton's mathematical sense, but in a more robust realist sense, we would evidently interpret Newton's contention that gravity is a universal force to mean that all bodies with mass attract one another across empty space, and we would construe this to involve the identification of a "physical cause or reason" for the motions of the massive bodies in question. Many of Newton's interpreters took this very idea to be incoherent; and Newton's detractors, such as Leibniz, attempted to foist this reading on Newton as the best construal of his own contentions. If we add the fact that Newton himself appeared to concur with Leibniz's sentiment in his famous correspondence with Bentley in 1692–3,[12] we obtain a powerful argument for the view that we ought to endorse the anti-metaphysical reading of Newton's mathematical treatment of force.

The anti-metaphysical interpretation can be supported further by a nuanced reading of Newton's most famous methodological pronouncement in the *Principia*, "*hypotheses non fingo*," "I feign no hypotheses."[13] Just as the mathematical treatment of force can be interpreted as expressing a strict causal agnosticism, focusing solely on empirically based descriptions of motions within the solar system, Newton's methodology can be interpreted as expressing a more general metaphysical agnosticism. *Hypotheses non fingo* might be interpreted to mean that any metaphysical view is a mere hypothesis, and therefore not to be endorsed in natural philosophy.[14] To

[11] The most extensive discussion of this problem in its historical context is Hesse, *Forces and Fields*, especially chs. 5, 6, and 7.

[12] See Newton's fourth letter to Bentley of 25 February 1693 in *Correspondence*, vol. III: 253–4.

[13] This was added to the second (1713) edition of the text. We owe this translation of the phrase to Alexandre Koyré, who first noted that Newton uses the word "feign" in a parallel discussion in English: *From the Closed World to the Infinite Universe*, 229 and 299 n. 12.

[14] For instance, consider Donald Rutherford's persuasive description of Newton's methodology:

> By the end of the seventeenth century, a distinction had begun to emerge between, on the one hand, natural science, characterized by experimentation, measurement, and mathematical representations of natural order, and, on the other hand, philosophy, conceived more or less traditionally as a speculative discipline. Although Newton's *magnum opus* is entitled *Mathematical Principles of Natural Philosophy*, Newton himself is one of the people most responsible for drawing a methodological boundary between natural science and speculative philosophy. The boundary is marked by his famous assertion "hypotheses non fingo" ("I do not feign hypotheses"). For Newton, the domain of science, or "experimental philosophy," is confined to explanatory propositions that can be "deduced from the phenomena." What cannot be deduced in this way is merely a hypothesis, and "hypotheses, whether metaphysical or physical, or based on occult qualities, or mechanical, have no place in experimental philosophy." (Rutherford, "Innovation and Orthodoxy in Early Modern Philosophy," 12)

> See also the discussion in Hanson, "Hypotheses Fingo" – according to Hanson's considered view, a hypothesis is simply "an expression of some philosophical or metaphysical prejudice" (32). Cf. Koyré, *Newtonian Studies*, 113.

explore this reading I will review four especially salient aspects of Newton's methodological pronouncement. As is well known Newton's attitude toward "hypotheses" shifts remarkably over time, and certainly may not be consistent through all of his writings even in a single period of his career.[15] The goal here is simply to emphasize several relevant aspects of Newton's conception of hypotheses during the time (1713) of the second edition of the *Principia* (and after) in order to indicate the textual support for the anti-metaphysical interpretation of Newton's work.

The first and most general aspect concerns the following question. In contending that he does not feign hypotheses, does Newton mean that he eschews hypotheses *per se*? This would be inaccurate, for Newton actually argues that within the boundaries of experimental philosophy – the *Principia* and the *Opticks* (excepting the queries) can be considered works in this area – one may not hypothesize, but it is not improper to propose hypotheses to prod future experimental research. Such hypothetical speculations are either reserved for the queries to the *Opticks*, or are more or less explicitly labeled as such in the *Principia*. For instance, in the Scholium to proposition ninety-six of book I of the *Principia*, Newton discusses hypotheses concerning light rays. Similarly, in query 21 of the *Opticks* he proposes that there might be an ether the differential density of which accounts for the gravitational force acting between bodies (I discuss that idea in greater depth in ch. 3).

The second aspect concerns the criterion that renders a proposition a *hypothesis*. The case of the postulated ether in query 21 might guide us, for the most salient fact about the ether is that Newton lacks independent experimental evidence indicating its existence. This coheres with Cotes's rejection, in his preface to the *Principia*'s second edition, of the common hypothesis that planetary motion can be explained via vortices on the grounds that their existence does not enjoy independent empirical confirmation (*Principia*, 393). So hypotheses may make essential reference to entities whose existence lacks independent empirical support. With such support the explanation would successfully shake off the mantle of "hypothesis." For Newton and Cotes the contention might be that both Descartes and Leibniz proceed in a "hypothetical" manner by attempting to explain phenomena through invoking the existence of entities – such as vortices – for which there is no independent empirical evidence.[16]

---

[15] See especially Cohen, "Hypotheses in Newton's Philosophy." For an important discussion of the role of hypotheses in Cartesian natural philosophy, see Clarke, *Occult Powers and Hypotheses*, 141–63.

[16] Although there may certainly be debate concerning the proper interpretation of the phrase "independent empirical evidence," one should not infer from this fact that this type of concern always

The third aspect follows closely on the second. What precisely does it mean to feign some hypothesis, as opposed to adopting some other cognitive attitude toward it? The queries to the *Opticks* clearly indicate Newton's avowed willingness to consider all manner of hypotheses; the salient point – which Newton makes in the "Account," his extensive polemic against various aspects of Leibniz's thought in the early eighteenth century[17] – is that he explicitly separates the queries from the rest of the *Opticks* to avoid the charge that he has "feigned" the hypotheses within them. This highlights the subtlety of Newton's attitude toward hypotheses, which is easily missed. As we have seen, a proposition – for instance, "The motion of the planets in elliptical orbits around the sun is caused by the action of an ether with differential density at distinct points in space" – will be labeled a hypothesis if there is no, or at any rate obviously insufficient, independent empirical evidence indicating the existence of the postulated entity, in this case the ether. But that same proposition can be considered as a prod to further empirical research; it is not "feigned" unless one adopts an unwarranted epistemic attitude toward it, for instance, if one asserts it to be the correct explanation of some reasonably well-documented natural phenomenon. The queries, then, press us to distinguish the epistemic status of a proposition *vis-à-vis* a relevant body of empirical data, and the proper epistemic attitude toward such a proposition, given all the relevant empirical data. Newton does not feign hypotheses in the General Scholium to the *Principia* in order to present a causal explanation of gravity – for instance, he does not contend that gravity must be due to the operation of an ether – but he is certainly willing to speculate about the possible properties of an ether in query 21 to the *Opticks*, as he had already done at the end of his famous 1679 letter to Boyle.[18]

The fourth and final aspect of Newton's methodology is particularly salient because it involves this question: what type of view generates any

separated Newton from his interlocutors. In some cases, for instance, defenders of vortices may simply have been willing to proceed in a hypothetical manner – postulating the existence of vortices in the solar system – on the grounds that the mechanical philosophy demanded some type of contact action to account for the planetary orbits, regardless of the independent empirical evidence for vortices. For instance, in order to account for the motions of the planetary bodies in his *Tentamen*, Leibniz introduces *ex hypothesi* the premise that some kind of fluid surrounds, and is contiguous with, the various planetary bodies, and then argues that this fluid must be in motion. I discuss this case in more detail below. See the *Tentamen* in Leibniz, *Die mathematische Schriften*, vol. VI: 149, and Bertoloni Meli, *Equivalence and Priority*, 128–9.

[17] The most relevant sections (at least for our purposes here) of the "Account" are reprinted in *Philosophical Writings*, 123–6. The original was published as: "An Account of the Book Entitled *Commercium Epistolicum*," in the *Philosophical Transactions of the Royal Society*.

[18] For the letter to Boyle, see *Correspondence*, vol. II: 288–96. Of course, as I note in ch. 3, the two ethers differ from one another because of the intervening discovery made in the *Principia* regarding gravity.

given hypothesis in the first place? Since *hypotheses non fingo* concerns the postulation of a cause for gravity, one might proceed by examining the kind of characterization of gravity's cause that Newton took himself to be eschewing with his methodological pronouncement. And since Leibniz and Newton discussed the issue of gravity's cause in their (often ignored) correspondence of 1693, Leibniz's theory of gravity is an appropriate example.

The *Tentamen* of 1689 represents one of Leibniz's most sustained discussions of some of the issues treated in the *Principia*, especially certain questions concerning the planetary orbits.[19] In this text Leibniz emphasizes that before any empirical research is completed within physics, we know both the nature of motion and the nature of bodies; these not only constrain any empirical research, they help us to understand (e.g.) basic astronomical data as providing evidence for certain conclusions rather than others. In particular, the "nature of motion" is expressed by what can be called the principle of inertia, so that moving bodies tend to recede along the tangent to any curve. It is the "nature of bodies" to be such that the state of motion of any given body can be altered only by something "contiguous" to it. Hence it is part of the nature of bodies that there can be no action between distant bodies. If we begin from the assumption that the planets follow curvilinear paths around the sun, it follows from the nature of motion that something must intervene to prevent them from following the tangents to their orbits; and it follows from the nature of bodies that whatever alters their motion in this respect must be "contiguous" to them.

Leibniz's argument then proceeds as follows. We introduce, *ex hypothesi*, the claim that a fluid surrounds, and is contiguous to, the various planetary bodies, and we then make the following argument to prove that this imagined fluid must be in motion:

To tackle the matter itself, then, it can first of all be demonstrated that according to the laws of nature *all bodies which describe a curved line in a fluid are driven by the motion of the fluid*. For all bodies describing a curve endeavor to recede from it along the tangent (from the nature of motion), and it is therefore necessary that something should constrain them. There is, however, nothing contiguous except for the fluid (by hypothesis), and no conatus is constrained except by something contiguous in motion (from the nature of the body), therefore it is necessary that the fluid itself be in motion.[20]

---

[19] For discussions of some of Leibniz's views, see Bertoloni Meli, *Equivalence and Priority, passim*, and Koyré, *Newtonian Studies*, 124ff. Newton's notes on the *Tentamen* are available in *Correspondence*, vol. VI: 116–22.

[20] See Leibniz, *Die mathematische Schriften*, vol. VI: 149; I follow Bertoloni Meli, 128–9.

Proposing a causal analysis of the planetary orbits must involve an attribution of contact action between the planets and some physically characterized entity or medium – such as a vortex – that is contiguous to them.[21] Leibniz's strategy is clear: one cites a general metaphysical presupposition or requirement – taken, for instance, from a conception of the strictures on causal attributions imposed by the mechanical philosophy – and then employs that requirement or presupposition as a premise in an argument whose conclusion involves an existence claim, one not supported independently of that premise.

Several of Newton's more famous pronouncements concerning hypotheses in his later years indicate the appropriateness of the example of Leibniz's argument in the *Tentamen*. For instance, in the General Scholium, he contends: "For whatever is not deduced from the phenomena must be called a hypothesis; and hypotheses, whether metaphysical or physical, or based on occult qualities, or mechanical, have no place in experimental philosophy" (*Principia*, 943). In query 28 to the *Opticks*, when considering the view of certain ancients that gravity can be attributed to some cause "other than dense matter," Newton writes:

Later philosophers banish the consideration of such a cause out of natural philosophy, feigning hypotheses for explaining all things mechanically, and referring other causes to metaphysics: whereas the main business of natural philosophy is to argue from phenomena without feigning hypotheses, and to deduce causes from effects, till we come to the very first cause, which certainly is not mechanical. (*Opticks*, 369)

It therefore appears that hypotheses may be generated from various sorts of metaphysical principles or views, and so the exclusion of hypotheses may represent a way of distinguishing "experimental philosophy" (or "physics") from metaphysics.

Thus the anti-metaphysical reading of Newton's work is strongly supported both by a rather natural, and well-regarded, interpretation of his treatment of force, and by a sophisticated interpretation of *hypotheses non fingo* that helps to illuminate Newton's relation to his most important interlocutor, Leibniz.

But there is another benefit of the anti-metaphysical reading of Newton's work: it can also be employed to highlight a parallel between the method of the *Principia* and Newton's famous treatment of light in the papers published in the *Philosophical Transactions* in the early to mid-1670s.[22]

---

[21] In this case the physical characterization indicates that we are considering a massive swirling fluid surrounding the planets.

[22] For an especially helpful discussion of this issue, one that does not defend the interpretation under discussion here, see Harper and Smith, "Newton's New Way of Inquiry," 113–27.

Newton's "mathematical" treatment of gravity, and his attempt to refrain from a "physical" treatment of that force, resembles his attempt in optics to separate his experimental approach from accounts – such as Robert Hooke's – of the physical character of light. Newton contends that the rays of light are differentially refrangible, and after gleaning data from various experiments with prisms, he concludes that rays of sunlight, or of ordinary white light, contain within them a spectrum of colors that can be separated out through the use of a prism. Colors are "connate" properties of rays of sunlight; the prism does not create the spectrum of colors. These are certainly physical claims in an ordinary sense: some physical interaction between prisms and rays of sunlight results in a spectrum because the former separates out the immanent colored rays of the latter. Or so he concludes from his experimental data.

Yet the experimental data do not settle what we might characterize as the more fundamental physical question regarding light, one with an extensive historical pedigree. Rays of light are what we all see when sunlight shines through a window – the question, of course, is how to characterize the fundamental elements comprising the rays. Does light consist of particles, or of waves? (Hence rays of light are not to be conflated with waves.) If light is particulate, what are the relevant physical properties of the particles? If it is not, what are the relevant physical properties of the waves? And in general, what properties of light's constituents – whether particulate or wavelike – account for the differential refrangibility of the rays? The idea, then, is that we might liken what we do not know in this case to what we do not know in the case of the force of gravity – namely, the relevant physical properties of the objects that account for the force of gravity. To what extent is this potential parallel borne out by the debate that Newton's optical researches generated in the 1670s?

At the highest level of generality, I will interpret Newton's discussions of issues in optics in the early 1670s to indicate two points. First, from his earliest forays into scientific debate, Newton became embroiled in discussions of general characterizations of the proper aims and methods of the sciences. And second, many of these discussions of scientific method and goals hinged on the proper role of hypothesizing, and of hypotheses, in natural philosophy. In February of 1672 an article appeared in the *Philosophical Transactions* with the title: "A Letter of Mr. Isaac Newton."[23] In this discussion Newton attempts to distinguish the presentation of an empirically based scientific theory from

---

[23] See "A Letter of Mr. Isaac Newton," which is reprinted in *Isaac Newton's Papers and Letters on Natural Philosophy*. Newton's so-called second paper on light and colors, read at the Royal Society in 1676, is also reprinted in the edition by Cohen and Schofield. Cf. also *The Optical Papers of Isaac Newton*, vol. I.

the presentation of what he would later term "hypotheses" concerning the nature of phenomena described by a theory.

Consider the structure of Newton's argument in this letter.[24] Through a series of experiments with prisms Newton attempts to undermine what he takes to be the common assumption that the rays of sunlight are equally refrangible. In particular, he employs a two-stage experiment involving sunlight passing through two prisms in order to support his conclusion that light rays are differentially refrangible, eventually concluding that the differential refrangibility of the rays of light into which the sunlight is broken by the prism is correlated with the colors of those rays. Based on his further contention that light rays are differentially refrangible originally – i.e. this is not due to the alteration of the light by a prism – Newton concludes that sunlight consists of constituent colored rays of light; colors do not result from modifications of sunlight. The literature on this set of experiments, and the argument based on them, is robust, and I want to leave aside any potential problems with the experiment and the argument in order to focus on the insight that this early paper gives us into Newton's role in foundational debates.

Once his paper was published in the *Philosophical Transactions*, Newton immediately argued that his critics systematically misunderstood his "theory" of light and colors. Robert Hooke, one of Newton's more prominent critics, responded to the argument above with a detailed letter to Henry Oldenburg, the Royal Society's secretary. From Hooke's perspective Newton's "theory" or "hypothesis" did not principally concern claims about differential refrangibility and heterogeneity; rather, differential refrangibility and heterogeneity represented alleged properties of the phenomena that had to be saved by a theory or hypothesis.[25] So Hooke searched Newton's paper for such a hypothesis and found the notion, mentioned briefly by Newton, that light was a "body."[26] Hooke then took his debate with

---

[24] In general I endorse the persuasive interpretation in Sabra, *Theories of Light from Descartes to Newton*, 240ff. Cf. also the discussion in Blay, *Les "Principia" de Newton*, 34–8, which places Newton's early optical research in the larger context of his intellectual development and work in the *Principia*.

[25] That Hooke does not think of these issues as forming an essential part of Newton's theory is clear for two reasons: (1) he is not concerned to reject the claim about differential refrangibility, but is concerned to reject Newton's theory; and (2) he takes his own hypothesis to be capable of accounting for both the fact about refrangibility and the fact, if it is a fact, about heterogeneity. See Hooke to Oldenburg, 15 February 1672, in Newton, *Correspondence*, vol. I: 112 on the former point, and 113–14 on the latter.

[26] See Hooke to Oldenburg, 15 February 1672, in Newton, *Correspondence*, vol. I: 113. In recounting Newton's theory Hooke does mention the points about refrangibility and heterogeneity, but he thinks that Newton's "first proposition" is "that light is a body" and that differently colored rays of light are in fact "several sorts of bodies." I take this to represent Hooke's interpretation of how Newton can account for the data with the theory that light consists of particles.

Newton to hinge on whether light consisted of particles, as he thought Newton maintained, or of waves, as Hooke alleged. Hooke was certainly not alone in this interpretation; indeed, his reading prevailed even many years later. For instance, in a 1694 letter to Huygens explaining Newton's theory of light, Leibniz writes that Newton took light to be a "body" propelled from the sun to the earth which, according to Leibniz, Newton thought could explain both the differential refrangibility of rays of light and the phenomena of colors.[27]

So for Hooke, and possibly for Newton's other interlocutors, a scientific theory or hypothesis is, broadly speaking, a conception of the fundamental nature of some phenomenon. One accounts for the relevant empirical data – one "saves" the phenomena – precisely by describing this nature. This does not entail Hooke or others taking the saving of the phenomena in this sense to determine which hypothesis about the nature of light is correct; on the contrary, Hooke's principal point is that his theory saves the phenomena as well as Newton's does. The point is that from Hooke's perspective it makes little sense to claim that Newton's presentation of empirical data concerning the properties of light based on experiments with prisms could itself constitute a scientific theory, independent of some hypothesis concerning the fundamental nature of light.

What might unify the discussions in the 1670s concerning optics, and the problem of gravity's cause that Newton tackles in the 1713 edition of the *Principia*, is not the employment of the term "hypothesis," but more importantly a debate concerning the proper goals of scientific theorizing. It is not that Newton articulates a single consistent view of the role of hypotheses in natural philosophy; on the contrary, his view changes over time, as does his use of the term "hypothesis." Rather, he consistently tackles matters such as the role of hypotheses because he consistently discusses foundational issues within natural philosophy, including especially the question of what the proper goals and methods of this field should be. And in each case we can reasonably understand Newton as eschewing speculative matters – such as the wavelike or particulate character of light, or the underlying basis of gravity – by focusing attention solely on those issues

---

[27] Leibniz to Huygens, 26 April 1694, in Huygens, *Œuvres complètes*, vol. X: 602. Huygens's own brief response to Newton was published in the *Philosophical Transactions* in July of 1673 – see his "An Extract of a letter." Ignatius Pardies, another of Newton's interlocutors, similarly found it difficult to differentiate the claim about the corporeal nature of light from Newton's ideas concerning refrangibility and heterogeneity. See his two letters to the Royal Society concerning Newton's work, both of which are reprinted in *Isaac Newton's Papers and Letters*; cf. the discussion of Pardies in Sabra, *Theories of Light*, 264–7.

that are tractable through various types of empirical and mathematical work. Hence Newton should be understood as fundamentally severing the link between various speculative – including traditional metaphysical – issues and the development of empirical science. These aspects of Newton's work in optics and in physics provide substantial support for the anti-metaphysical interpretation, and may account for its popularity in the eighteenth century.

NEWTON'S RADICAL EMPIRICISM

Beginning in 1967, and continuing with a remarkable series of papers in the intervening years, Howard Stein has developed a powerful interpretation that differs fundamentally from the anti-metaphysical reading of Newton presented above.[28] From Stein's point of view, which has also been endorsed in crucial respects by Robert DiSalle,[29] Newton does not reject a Cartesian understanding of the metaphysical foundations required for physics by eschewing metaphysical concerns altogether; rather, he follows the quasi-Aristotelian view that metaphysics ought to "come after" physics. In Stein's eyes Newton thinks that without the guide of secure results in physical theory, metaphysical work would consist solely of speculation and hypothesizing. Stein's Newton is indeed silent on a number of pressing metaphysical issues, but only because the best physical theory of the day – the theory presented in the *Principia* – provided no clear guide on how to address those issues. When the "secure results" of empirical scientific research can serve as his guide Newton is willing to address fundamental questions about space, time, motion, and matter. For Stein and DiSalle, Newton not only attempts to constrain metaphysics with the results of the physical sciences – he takes physics to be logically prior to metaphysics, transforming metaphysics from an *a priori* investigation of the structure of the world into a thoroughly empirical investigation that can make progress only by looking to the development of empirical science.

To see what might motivate what I call the radical empiricist interpretation in part endorsed by Stein and DiSalle, recall that the anti-metaphysical reading of Newton's work sketched above is supported both by a well-regarded interpretation of his "mathematical" treatment of force and a

---

[28] Stein's first paper on Newton that addresses these issues was given as a talk at a conference in 1967, and subsequently published, with some revisions, in 1970 as "Newtonian Space–Time." Several of his other papers are listed in the bibliography.

[29] DiSalle, "Newton's Philosophical Analysis of Space and Time" and *Understanding Space–Time*.

sophisticated interpretation of *hypotheses non fingo* that illuminates Newton's relation to Leibniz. One way of understanding the motivation for the radical empiricist interpretation lies in recognizing the surprising intersection of the mathematical treatment of force with Newton's famous methodological pronouncement in the General Scholium to the *Principia*, an intersection that highlights further nuances.[30]

The best way to analyze the intersection of these two aspects of Newton's thought is to return to the context of *hypotheses non fingo*. Since the pronouncement occurs in the course of considering the cause of gravity, as we have seen, we ought to emphasize the conception of gravity's cause that Newton refrains from endorsing. But to understand the contrast that Newton has in mind, we should also emphasize the claims concerning gravity that he does endorse. In the General Scholium, we read:

> Thus far I have explained the phenomena of the heavens and of our sea by the force of gravity [*Hactenus phænomena cælorum & maris nostri per vim gravitatis exposui*], but I have not yet assigned a cause to gravity. Indeed, this force arises from some cause that penetrates as far as the centers of the sun and the planets without any diminution of its power to act, and that acts not in proportion to the quantity of the *surfaces* of the particles on which it acts (as mechanical causes are wont to do) but in proportion to the quantity of *solid* matter, and whose action is extended everywhere to immense distances, always decreasing as the squares of the distances . . . I have not been able to deduce from phenomena the reason for these properties of gravity, and I do not feign hypotheses [*hypotheses non fingo*] . . . and it is enough that gravity really exists [*Et satis est quod gravitas revera existat*], acts according to the laws that we have set forth, and suffices for all the motions of the heavenly bodies and of our sea.[31]

Immediately before proclaiming that he does not feign any hypothesis concerning the cause of gravity – for instance, he does not feign the vortex hypothesis favored by Leibniz – Newton presents the astonishing claim that gravity does not act as "mechanical causes are wont to do." Any understanding of this claim obviously must cohere with an understanding of Newton's methodology – surely, gravity's non-mechanical character is not by hypothesis, according to Newton – but it must also cohere with our best understanding of the particular method in the *Principia* that supports this final, startling claim. And that method is what Newton calls the

---

[30] To be clear, the first goal of this section is merely to further the dialectic of this chapter by indicating a difficulty with the interpretation of Newton offered in the first section; this is intended, in turn, to motivate the Stein/DiSalle reading. But none of what I say in this first part of this section about hypotheses and the General Scholium ought to be attributed to Stein and DiSalle. I tackle their own readings of Newton explicitly in the next part of this section.
[31] *Principia Mathematica*, vol. II: 764. For discussion, see ch. 3.

"mathematical treatment of force." So an interpretation of Newton's methodological pronouncement ought to be presented in conjunction with an interpretation of the general method in the *Principia* regarding force, for these two aspects of Newton's work intersect.

Newton's distinction in the *Principia* between the mathematical and the physical treatment of force has three aspects that should illuminate the surprising contention that gravity is not mechanical.[32] Each of these three aspects is expressed in the passage from the General Scholium quoted above. First, the mathematical treatment of gravity determines the salient physical variables – mass and distance – in gravitational interactions, even while remaining silent on gravity's "basis." By that basis, Newton means the entity, or entities, that mediate gravitational interactions, whether they might be particles, the ether, or even an immaterial medium of some kind. Whatever particle or medium mediates such interactions, it does so in a way that reflects the law of universal gravitation. Second, since gravity is proportional to mass, but not to volume or surface area, it is non-mechanical in Newton's sense. That is, it acts not on the surfaces of bodies, but on all of their parts. Third, these contentions, in turn, express a commitment to a causal unification of previously disparate phenomena: for instance, free fall and the planetary orbits have the same cause. It is precisely this claim that is signaled by Newton's use of the term gravity to refer to this force, since gravity was previously understood to be a purely terrestrial phenomenon. Crucially, Newton contends that our ignorance of gravity's physical basis does not extend to the question of whether it operates "mechanically"; the proportionality to mass indicates that it does not. Therefore, we already know that some causation in nature is not mechanical in a substantial sense.

So the methodological slogan of the *Principia*'s third book is not incidentally, but in fact quite intimately, connected to the conclusions regarding force supported by its mathematical treatment. From Newton's point of view it is obviously not a hypothesis that gravity is non-mechanical; that is to say, he takes this conclusion to be supported by various kinds of empirical evidence. But he also admits that he does not know the "reason" for the "properties" of gravity expressed in the law of universal gravitation, which is to say that he lacks a complete physical treatment of this force. We do not know what it is about the ether, or the exchange of particles, etc., that

---

[32] My discussion of the mathematical treatment of force should not be attributed to Stein and DiSalle; it is intended to provide an additional motivation for their reading of Newton, which I then describe in more detail in this section below. In chs. 3 and 6, I discuss further nuances to Newton's claim that gravity is not mechanical – it might remain mechanical in other senses that Newton does not mention here, such as the fact that it is governed by the laws of motion. These issues can be bracketed here.

renders mass and distance the salient variables in the relevant causal interactions. Similarly we do not know what it is about this physical basis that
explains how it could possibly be that the planetary orbits, free fall, the tides,
etc., all have the same cause. And he refuses to feign any hypothesis in lieu of
that physical treatment – i.e. he refuses to hypothesize what the physical
basis of gravity might turn out to be. In this sense the methodological slogan
is perfectly consistent with the endorsement of a strong and explicit metaphysical contention, and indeed, the very contention that placed Newton at
the center of a heated controversy concerning force that occupied much of
the eighteenth century. He had claimed that some causation in nature was
non-mechanical.

This last point can be put in stronger terms. If we consider the intersection of the two striking claims from the passage in the General Scholium
quoted above – gravity operates non-mechanically and *hypotheses non fingo* –
we find that each of them represents a response to the mechanical philosophy. As is generally acknowledged, in presenting his methodological
slogan Newton asserts that Leibniz is wrong to insist on a mechanist
conception of gravity; for his part, Newton will not feign any such hypothesis. But here is the crucial point: that is not because Newton rejects the
metaphysical presupposition that all causation must be mechanical on
methodological grounds. More radically he explicitly contends that some
causation in nature is non-mechanical and he does so based on physical
theory itself.

This episode leads rather naturally to a substantial modification of the
anti-metaphysical reading of Newton presented above, for we have found
that Newton explicitly challenges the then-prevailing mechanical philosophy, reaching the surprising conclusion that some causation in the natural
world is non-mechanical. This represents, of course, a radical rejection of
the view held by many of Newton's predecessors, such as Descartes, and by
many of his critics and interlocutors, such as Leibniz and Huygens. In
this context Newton emerges as a figure willing to engage with the most
prevalent metaphysical conception of his day, rejecting it in a fundamental way.[33] Thus Newton does not eschew the metaphysical questions
that animate the philosophical debates of his day; he provides answers to
those very questions, but from a surprising, if not unique, source: the

---

[33] As noted above, my use of "metaphysical" and of "metaphysics" is intended to be as neutral as possible
between the three interpretations of Newton characterized in this chapter. Certainly the mechanical
philosophy and its characteristic issues involved metaphysical questions from the perspective of some of
its adherents and critics, and these issues are metaphysical in a straightforward sense from a contemporary point of view as well. Thanks to Karen Detlefsen for discussion of this point.

development of empirical science itself. This indicates why one might conclude that Stein and DiSalle are fundamentally correct in taking Newton to endorse a radically empirical perspective on the questions that drive some pressing metaphysical discussions in the late seventeenth century. For Newton does not merely reject the mechanical philosophy of Descartes, Leibniz, and Huygens; he transforms what they take to be a purely *a priori* question – what kinds of causation exist in the natural world at the most basic level? – into an empirical question, one answerable only through reflection on the development of physics.

It is this transformation of previously *a priori* questions into empirical ones that serves as a hallmark of the interpretation defended by Stein and DiSalle.[34] In their eyes Newton is perfectly willing to endorse positions on various questions that are obviously metaphysical in some sense – positions concerning, for instance, the structure of space and time, or the kinds of causation found in the natural world – but only after a fundamental rejection of the Cartesian, *a priori* approach to such questions has been effected. Stein and DiSalle's Newton inherits a series of obviously metaphysical questions from the natural philosophical tradition represented by figures such as Descartes, but he transforms the consideration of those questions by taking the secure results of physical theory to dictate his answers. Hence in Stein's words, Newton revives the "Aristotelian" view that physics ought to precede – perhaps in the sense of being logically prior to – metaphysics. The result is not an anti-metaphysical Newton but a kind of empiricist metaphysician.

How do Stein and DiSalle approach other areas of Newton's work? Whereas the anti-metaphysical reading of Newton outlined above tends to characterize Newton's discussion of space and time – for instance, in the Scholium to the *Principia* – as a mere residue of outmoded metaphysical thinking that plays no role in Newton's physical theory, Stein's first, and perhaps most influential, innovation was to argue instead that the Scholium on space and time in the *Principia* indicates that the distinction between absolute and relative space, time, and motion is supported by the understanding of motion articulated in Newton's laws. Stein had powerfully argued already in his 1970 paper, "Newtonian Space–Time," that we must interpret the Scholium as presenting an alternative to the conception

---

[34] The meaning of the distinction between *a priori* and *a posteriori* has obviously shifted over time, and is particularly different in the pre- and post-Kantian eras. To reflect what I take to be the perspective adopted by Stein and DiSalle, I intend an *a priori* proposition to be one that is justified (or justifiable) independent of any appeal to experience, and an *a posteriori* proposition to require such an appeal. Thanks to Karen Detlefsen for raising this issue.

of space, time, and motion articulated by Descartes in part II of his *Principia Philosophiae*. Far from attempting to rebut a "relationalist" position later associated with Leibniz, Newton attempted to indicate that we should not understand the true motion of a body as consisting in a change in its relations with other bodies, as Descartes did. In order to understand true motion in a way that is consistent with his laws, Newton postulates absolute or mathematical space, allowing him to conceive of true motion as a change in absolute place. This move also enables Newton to save the perceptible effects of accelerating bodies – most famously noted in the examples of the rotating bucket and the connected globes in the Scholium – since all accelerations can be understood as true motions within absolute space.[35]

But Stein did not merely, and very influentially, argue that the Scholium on space and time captured facts about motion that were expressed in Newton's three laws; in subsequent work, culminating in "Newton's Metaphysics" (2002), he addressed Newton's other reflections on space in the General Scholium and in *De Gravitatione*, arguing that his views throughout should be understood as fundamentally empirical in character. Thus from Stein's point of view, Newton's well-known contention that God occupies every place within infinite space at every instant, and his controversial view that even the human mind occupies places within space (*contra* Descartes, of course), are each empirical in character. They must be interpreted as following from our ordinary perceptual experience of objects in the world, and of our own bodies, and perhaps from the interpretation of various claims within sacred texts, such as the Bible and the works of the prophets.[36] The resulting picture, then, is of a Newton whose rejection of Cartesian metaphysical foundations for physics and empirical research is so thoroughgoing that he rejects the very idea of *a priori* knowledge, even of the most basic facts about the divine being. In this way Stein's Newton inherits concepts from the late seventeenth-century metaphysical tradition, such as that of the necessary being, but then adopts the radical view that our knowledge of such matters is purely empirical.

---

[35] Although all accelerations are true motions, not all true motions are accelerations, since any change of absolute place is a true motion. Thus a well-known tension can be generated between Newton's laws and their corollaries on the one hand, and the conception of absolute space in the Scholium on the other. Whereas the notion of absolute space gives rise to the idea of the true velocity of every body, Newton's fifth corollary to the laws of motion indicates that the true velocity of a body can never be measured; only its accelerations can be measured (*Principia*, 423). Then in corollary six Newton indicates further that if every part of a system of bodies is accelerated identically, then even the acceleration of those bodies cannot be detected (*Principia*, 423). I discuss this issue in greater detail in ch. 5.

[36] These, too, are sources of empirical knowledge, for Stein's Newton – see ch. 5 for details.

In a series of papers culminating in *Understanding Space–Time* (2006), DiSalle adds a crucial element to Stein's overarching interpretation, one focused on a question that plagued Newton after the publication of the *Principia* in 1687. According to DiSalle, Newton's interlocutors and contemporaries took the notion of action at a distance to contradict the "very idea" of physical action, and therefore felt compelled to reject Newton's theory of gravity if it postulated such action. But to DiSalle's Newton we cannot give an *a priori* analysis of the concept of action, or of causation more generally; rather, action between material bodies must be defined solely through the laws of motion, which provide empirical criteria for measuring the action of one thing on another.[37] And so as DiSalle puts it, "Thus the question of action at distance became an empirical question."[38] For Newton "any interaction is physically intelligible as long as, and just to the extent that, it conforms to the laws of motion."[39] Since it is a merely empirical question Newton's theory of gravity can be understood as indicating, for the first time, that in fact the bodies within the solar system do causally interact with one another at a distance.[40]

In this crucial sense, then, DiSalle takes a remarkable step beyond the point reached in the discussion of non-mechanical causation above, arguing that Newton thinks of causation and even the vexing issue of action at a distance as involving fundamentally empirical questions. In that regard DiSalle's Newton effects an even more fundamental rejection of the

---

[37] It is a non-trivial fact that Newton's laws are in fact perfectly consistent with states of affairs in which material bodies act at a distance on one another (cf. the discussion of the third law in ch. 6). In that regard, DiSalle's interpretation is on solid ground. This fact is non-trivial because at least one of Newton's significant predecessors, Descartes, articulates laws of nature that appear to rule out action at a distance by requiring contact action alone for any deviation from the inertial state of a material body. See the helpful analysis in Suppes, "Descartes and the Problem of Action at a Distance," 149–50.

[38] See DiSalle, "Newton's Philosophical Analysis of Space and Time," 52; see also *Understanding Space–Time*, 42 and 72.

[39] DiSalle, *Understanding Space–Time*, 42.

[40] DiSalle is certainly not alone in contending that – at least in some sense – Newton accepts the idea that natural change may occur through action at a distance among material bodies. For instance, see Westfall, *Force in Newton's Physics*, 395 and McMullin, *Newton on Matter and Activity*, 150 n. 210; cf. also Heilbron, *Elements of Early Modern Physics*, 44–6. The argument against DiSalle that I present here will presumably also hold for their interpretations. As for Stein, his position on this issue is not entirely clear. On the one hand he apparently interprets Newton as holding that all of our knowledge, including even our knowledge of God, is empirical in character – this may imply that Newton's claim that action at a distance is "inconceivable" is in fact an empirical one, and therefore subject to revision under some conditions. On the other hand, although Stein takes Newton to believe that even God is a local actor ("Newton's Metaphysics," 270), he apparently does not address the status of Newton's rejection of distant action directly. DiSalle's originality may lie in his characterization of Newton's transformation of the question of action within the physical world into a purely empirical question, which I take to follow the spirit, if not the letter, of Stein's work.

mechanical philosophy than that articulated by Stein. On this picture
Newton's rejection of mechanism is twofold. First, he contends that some
causation in nature is not mechanical, since gravity is proportional to mass.
Second, he does not proclaim his ignorance of gravity's basis; instead, he
contends that gravity has no basis because it involves a direct action between
material bodies across empty space. Thus DiSalle's Newton attributes to
bodies a kind of action that thinkers such as Leibniz and Huygens take to be
impossible on purely *a priori* grounds. In this way the picture of Newton
that emerges in DiSalle's work serves as a crucial complement to the view
provided by Stein.

The radical empiricist interpretation defended in large measure by Stein
and DiSalle serves as an important corrective to the anti-metaphysical
reading of Newton that first gained prominence in the early to mid-
eighteenth century. In lieu of attributing a general metaphysical agnosticism
to Newton – one that has difficulty, for example, in understanding the
remarks on space and time as anything but a leftover from earlier ways of
thinking – we attribute to him a principled empiricist attitude toward
metaphysical questions. Instead of engaging in idle speculation regarding
the basic structure of space and time, or the types of action that exist within
the natural world, Newton indicates how previously *a priori* metaphysical
issues can in fact be dealt with in a systematic and fruitful way by looking
toward the development of empirical science itself. And if empirical science
is silent on the many remaining topics within traditional seventeenth-
century metaphysics, then the natural philosopher must also remain silent:
hypotheses non fingo.

The fundamental question with which the radical empiricist interpretation leaves us, then, is whether we can in fact understand Newton's treatment of God's relation to space, and of action at a distance, as thoroughly
empirical in character. As we shall see the answer to this question lies in
recognizing the surprising intersection of Newton's conception of the
divine being on the one hand, and of the always prevalent question of
action at a distance on the other hand. For Newton any conception of
action must in part reflect our understanding of God's own action within
space and time – this will become central to the discussion below.

A PHYSICAL METAPHYSICS: INVERTING DESCARTES?

What precisely is action at a distance, as Newton considers it? The key to
answering this question lies in avoiding an anachronistic trap: in the twentieth
century, Einstein famously raised the issue of "spooky action at a distance"

within quantum mechanics after his special theory of relativity had set an upper limit on the transmission of information between spatially separated physical systems.[41] In this context physical systems can be separated by a spatial distance, but also by what we might call a temporal distance: two systems cannot interact within a given time interval if the interaction requires any transmission of information faster than the velocity of light. So two particles might, at least apparently, exhibit action at a distance on one another if their interaction appeared to occur faster than light's velocity. But the discussion of action at a distance in the late seventeenth century is entirely different: Newton and his interlocutors lacked our idea that spatially separated systems cannot interact instantaneously. Rather, they understood bodies to be distant from one another only if they were spatially distant. Hence on the classic mechanical model in optics articulated by Descartes, light is a pressure transmitted instantaneously through a medium.[42] If gravity were understood to be such a pressure – transmitted, say, through some kind of ether – then Jupiter and Saturn could interact instantaneously through the medium without acting at a distance.[43] This would of course become unacceptable in the twentieth century.

Newton's theory of gravity in the *Principia*, of course, is perfectly consistent with the contention that material bodies, such as the planets, act on one another directly across empty space – this might even be the most natural interpretation of the theory, and it was prevalent in the eighteenth century.[44] For Newtonian gravity acts instantaneously, and the *Principia* posits no medium for gravitational interactions. And yet Newton himself appears to reject action at a distance.[45] The most prominent discussion of

---

[41] Einstein's remarks were made after the special theory of relativity had established that physically separated systems cannot exchange information faster than the limit set by light's invariant velocity. And of course on one interpretation, it seems that sub-atomic particles emitted from a common source do in fact exchange information faster than that limit. Cf. also the helpful discussion of nineteenth-century developments in Hesse, "Comment on Howard Stein," 298–9.

[42] For a comprehensive historical analysis of various conceptions of light's velocity, see Cohen, "Roemer and the First Determination of the Velocity of Light (1676)" – Cohen discusses Descartes's views in depth at 333–7. Newton, of course, was well aware that light travels with a finite velocity; he was even in a position to estimate that sunlight reaches the earth in roughly seven or eight minutes (*Opticks*, 351). But he was not in a position to think of light's velocity as invariant – he understood it to be subject to ordinary velocity addition, as did his contemporaries – or to think of it as imposing an upper limit on the transmission of information between spatially separated systems.

[43] For an excellent discussion of related issues, including the question of how we should understand spatiotemporal locality, see Lange, *An Introduction to the Philosophy of Physics*, 1–26 especially. See also the illuminating discussion in Maxwell's entry on action at a distance in *The Scientific Papers of James Clerk Maxwell*, vol. II: 311–23.

[44] I discuss this – and several related – points in ch. 6.

[45] See the discussion in Buchdahl, "Explanation and Gravity," especially 176 n. 34, and in Boas, "The Mechanical Philosophy," 516–17.

the idea that material bodies might interact at a distance appears in a passage
from Newton's correspondence with Richard Bentley in 1693, as Bentley
was preparing to publish his Boyle Lectures:[46]

It is inconceivable that inanimate brute matter should, without the mediation of
something else which is not material, operate upon and affect other matter without
mutual contact, as it must be, if gravitation in the sense of Epicurus, be essential
and inherent in it. And this is one reason why I desired you would not ascribe
innate gravity to me. That gravity should be innate, inherent, and essential to
matter, so that one body may act upon another at a distance through a vacuum,
without the mediation of anything else, by and through which their action and
force may be conveyed from one to another, is to me so great an absurdity that I
believe no man who has in philosophical matters a competent faculty of thinking
can ever fall into it. (*Correspondence*, vol. III: 253–4)[47]

Prima facie it is unlikely that Newton would declare action at a distance
"inconceivable" in 1693 if his own theory of 1687 indicated that bodies

---

[46] For a discussion of the Boyle Lectures and their importance for spreading various "Newtonian" ideas,
see Jacob, "Christianity and the Newtonian Worldview," 243–6. Newton himself may have suggested
that Bentley deliver the lectures, which were endowed by Robert Boyle's will (Boyle died at the end of
1691). See the discussion in Koyré, *Newtonian Studies*, 201–2. For an indication of the influence
enjoyed by Bentley's lectures, including their appearance in several editions on the Continent, see
Metzger, *Attraction universelle*, part 2: 92 n. 2; she also provides an extensive analysis of the lectures in
Attraction universelle, part 2: 80–91.

[47] Newton's claims here are not in tension with one another, as is sometimes thought. Whereas he
informs Bentley that the medium in question is not "material," he indicates that in the *Principia* itself
he leaves open the question of whether there is any medium for gravitational interactions, and of
whether any medium must be material.

   Unfortunately, Bentley's letters to Newton, save one, are not available (they are not printed in
Correspondence or in *The Correspondence of Richard Bentley*). However, the copy of Newton's first
letter to Bentley (sent on 10 December 1692) in the library of Trinity College, Cambridge has a note
written in Bentley's hand that reads: "Mr Newton's answer to some queries sent by me after I had
preached my last two sermons [i.e. the Boyle Lectures]" – *Correspondence*, vol. III: 238. Thus Bentley
sent Newton a series of questions while he was preparing to publish his Boyle Lectures; Newton
eventually replied with a set of four letters to Bentley. Newton's letter quoted in the main body of the
text was in response to a long set of queries that Bentley sent Newton in a single letter on 18 February
1693 – *Correspondence*, vol. III: 246–53. Bentley discusses the issue Newton mentions in his eighth
Boyle Lecture, given in December of 1692, under the general title *A Confutation of Atheism from the
Origin and Frame of the World*, reprinted in *Isaac Newton's Papers and Letters*, 332–3, 338–45. Bentley's
formulation in his published lectures closely parallels the remarks in Newton's letter: "'Tis utterly
unconceivable, that inanimate brute matter (without the mediation of some immaterial being)
should operate upon and affect other matter without mutual contact; that distant bodies should
act upon each other through a *vacuum* without the intervention of something else by and through
which the action may be conveyed from one to the other" (*Isaac Newton's Papers and Letters*, 340–1).
Bentley's last two lectures may represent the first popular presentation of Newton's work in the
Principia. Newton was in contact with Bentley in the summer of 1691 as well, providing him with a
famous list of materials to read in preparation for understanding the *Principia* – see *Correspondence*,
vol. III: 154–5.

exhibited precisely this type of action in their gravitational interactions.[48] Moreover, the claim that gravity is not essential to matter – which Newton apparently takes to entail the claim that material bodies act at a distance on one another – also appears in the 1713 and 1726 (the last) editions of the *Principia*, in the discussion of the third rule of philosophizing (*Principia*, 795–6).[49]

These texts certainly call DiSalle's remarks into question. Newton endorses a view of action defended by nearly all of his contemporaries, including the Aristotelians, the mechanists, and those, such as Leibniz, who attempt to embrace mechanism while reviving aspects of Aristotelianism. As we will see in greater detail below, Newton held the familiar view that a substance cannot act where it is not. However, the Bentley letter was not published during Newton's lifetime – it first appeared in an edition in the mid-eighteenth century[50] – and Newton's public denial that gravity could be essential to matter does not itself foreclose the possibility that material bodies might exhibit distant action.[51] So DiSalle's view might still be

[48] According to McMullin, Newton's rejection of distant action among material bodies should not be interpreted as a rejection of action at a distance *per se* – see *Newton on Matter and Activity*, 144 n. 13 and 151 n. 210. To my knowledge McMullin does not address the question of God's action in the world; I do so below.

[49] Commentators sometimes contend that in a passage from query 31 to the *Opticks* (added to the second, 1706, Latin edition of the text, obtaining its numbering from the 1717 English edition), Newton actually embraces the possibility of action at a distance. But in fact, if one reads the full passage, one sees that Newton is repeating precisely the kind of point he makes about the *Principia*'s mathematical treatment of force in the Scholium following proposition 69 in book I. In the beginning of query 31, Newton writes:

Have not the small particles of bodies certain powers, virtues, or forces, by which they act at a distance, not only upon the rays of light for reflecting, refracting, and inflecting them, but also upon one another for producing a great part of the phenomena of nature? For it's well known, that bodies act one upon another by the attractions of gravity, magnetism, and electricity; and these instances show the tenor and course of nature, and make it not improbable but that there may be more attractive powers than these. For nature is very consonant and conformable to her self. How these attractions may be performed, I do not here consider. What I call attraction may be performed by impulse, or by some other means unknown to me. I use that word here to signify only in general any force by which bodies tend towards one another, whatsoever be the cause. For we must learn from the phenomena what are the laws and properties of the attraction, before we enquire the cause by which the attraction is performed. (*Opticks*, 375–6).

This passage raises a number of important questions, including the question of how we ought to interpret Newton's mathematical treatment of force in the *Principia* – I discuss that issue in depth in ch. 3.

[50] Newton's letters first appeared publicly in *Four Letters from Sir Isaac Newton to Doctor Bentley* in 1756.

[51] I take it that for Newton the claim that gravity is essential to matter entails the claim that material bodies act on one another at a distance, but not vice versa. As for the former: if gravity is essential to matter, then two material bodies would bear the property of gravity even at a spatial distance from one another, and even if they were the only two bodies in their world (if the property were to disappear

considered persuasive. Yet in addition to the interpretation of Newton's theory, and of his scattered explicit remarks, there is another crucial element here that merits consideration. Since Newton thought that substances could not act except where they were present, the question of how he understood the infinite substance's action becomes pressing. This question is not foreign to his work in natural philosophy: he declares in the General Scholium to the *Principia*, added to the second (1713) edition of the text, that "to treat of God from phenomena is certainly a part of natural philosophy."[52] And it seems to me that his conception of the divine being's place within the understanding of the physical world developed by the natural philosophy of the *Principia* suggests a fundamentally different picture than that offered by DiSalle.[53]

In both published and unpublished texts spanning many decades, Newton defends a heterodox conception of God's relation to space, time, and the physical world, one that most famously appears in the General Scholium, added to the second (1713) edition of the *Principia*.[54] In that text, he writes:

with the disappearance of other bodies, or with an increase in spatial separation between the bearers, it would be merely accidental). But if bodies bear the property of gravity in that situation, then *ipso facto* bodies act on one another at a distance. However, as for the latter: if bodies act on one another at a distance, that does not entail that gravity is essential to matter, for two otherwise lonely bodies at some spatial separation from one another might act in that fashion, but lose their property of gravity through some arbitrary increase in their spatial separation, thereby rendering the property merely accidental. (Newton himself notes that gravity cannot be essential to matter because it decreases with an increase in spatial separation – *Principia*, 796.) Thus Newton's denial in the *Principia* that gravity is essential to matter does not itself show that he denies action at a distance *per se*. For a helpful discussion of Newton's denial that gravity is essential to matter, see Harper, "Reasoning from Phenomena," esp. 168. Thanks to Eric Schliesser for help with these issues.

52   In the second edition of 1713 Newton writes that "to treat of God from phenomena is certainly a part of experimental philosophy" – he changed "experimental" to "natural" in the third edition of 1726 (see *Principia*, 943).

53   In a provocative essay John Henry has defended a view that bears similarities to DiSalle's interpretation, although on different grounds. From Henry's point of view, Newton should be understood as ultimately endorsing action at a distance. He writes: "I believe it is safe to say that all of Newton's pronouncements upon gravitational attraction are consistent with the view that he believed gravity to be a superadded inherent property of body which was capable of acting at a distance" (Henry, "Pray do not ascribe that notion to me," 141). However, Henry's conclusion contradicts the letter to Bentley quoted above, in which Newton denies, among other things, that gravity is "inherent" to matter. Henry also cites no textual evidence from Newton's published or unpublished work; instead, his evidence consists only of comments from Bentley, Clarke, and Cotes. But, as I suggest below, Newton's views should not be conflated with those of his followers.

54   Although it was decidedly controversial, some prominent Newtonians were willing to embrace and defend this view – see, for instance, Maclaurin, *An Account*, book IV: 384–5. Clarke's attempted defense of Newton in the face of Leibniz's criticisms proved more complex, as I discuss in depth in ch. 5.

He is eternal and infinite, omnipotent and omniscient, that is, he endures from eternity to eternity, and he is present from infinity to infinity; He is not eternity and infinity, but eternal and infinite; he is not duration and space, but he endures and is present. He endures always and is present everywhere, and by existing always and everywhere he constitutes duration and space. Since each and every particle of space is *always*, and each and every indivisible moment of duration is *everywhere*, certainly the maker and lord of all things will not be *never* or *nowhere*. (*Principia*, 941)[55]

Newton then adds: "It is agreed that the supreme God necessarily exists, and by the same necessity he is *always* and *everywhere*" (*Principia*, 942). Thus God is the first cause because God creates the physical world. But in a move away from a traditional picture of God's relation to the created world, the divine being occupies all places within space at all times, throughout the history of the world.[56] And this is necessarily the case, since it represents a consequence of God's eternity and infinity: for God to be an infinite being that exists forever just is for God to occupy all spaces at all times.[57] Newton dramatically emphasizes this view in a famous passage from the queries to the *Opticks*:

[Does] it not appear from phenomena that there is a being incorporeal, living, intelligent, omnipresent, who in infinite space, as it were in his sensory, sees the things themselves intimately, and thoroughly perceives them, and comprehends them wholly by their immediate presence to himself: of which things the images only carried through the organs of sense into our little sensoriums, are there seen and beheld by that which in us perceives and thinks. And though every true step made in this philosophy brings us not immediately to the knowledge of the first cause, yet it brings us nearer to it, and on that account is to be highly valued. (*Opticks*, 370)[58]

Thus God is never distant from any object at any time. God is present to all objects at all moments of time, throughout eternity and throughout infinite

---

[55] We find this same conception articulated in depth in *De Gravitatione* (e.g. at 25–6).

[56] Newton's critics often rejected this view out of hand – see, e.g., Leibniz's letter to J. Bernoulli in December 1715, where he calls the view "astonishing" (*Correspondence*, vol. VI: 260–1). In a subsequent letter to Bernoulli (27 May 1716), Leibniz contends that "space is the idol of the English [*spatium hodie est Idolum Anglorum*]" – *Correspondence*, vol. VI: 354, 356.

[57] I discuss these issues in much greater depth in ch. 5.

[58] Newton's notion that God sees objects "as it were in his sensory" led to a significant controversy with Leibniz, who objected to the idea on various grounds – see, for instance, section 3 of Leibniz's first letter to Clarke, and the third section of Clarke's first reply (Leibniz, *Die philosophischen Schriften*, vol. VII: 352–3). In a letter to J. Bernoulli in March of 1715, Leibniz writes that he laughed when he encountered this view – *Correspondence*, vol. VI: 213. Maclaurin discusses the episode in *An Account*, book IV: 383–4. For an important historical discussion of this issue, see Cohen and Koyré, "The Case of the Missing *Tanquam*"; see also the brief, but helpful, comment in Bloch, *La Philosophie de Newton*, 513–14.

space. From Newton's point of view, then, God in fact never acts at a distance on any object, at any time in the history of the world.[59]

Newton does not merely take God to be a local actor because God is never distant from any objects within the world – he even contemplates the consequences of contending that God is in fact distant from objects within the world. Newton discusses this issue explicitly in his (anonymous) "Account" of the Royal Society's report on the calculus priority dispute with Leibniz, published in 1715. Whether he accurately portrays Leibniz's views in such a text is obviously an open question, but it is useful for determining Newton's own preferred interpretation of his work in natural philosophy:

> The one teaches that God (the God in whom we live and move and have our being) is omnipresent, but not as a soul of the world: the other that he is not the soul of the world, but INTELLIGENTIA SUPRAMUNDANA, an intelligence above the bounds of the world; whence it seems to follow that he cannot do anything within the bounds of the world, except by an incredible miracle. (*Philosophical Writings*, 125)[60]

If God existed outside the spatiotemporal bounds of the physical world and was therefore distant from some object within the world, Newton argues, any causal influence that God exhibited on that object would involve "an incredible miracle." Given the context of this contention, the most likely

---

[59] In that regard Newton held a rather traditional conception according to which action must be local – this view has an Aristotelian pedigree, and was discussed frequently in the medieval period, for instance by Aquinas and Suarez. See especially Hesse, *Forces and Fields, passim*. Cf. also Bloch, *La Philosophie de Newton*, 512–13. Newton's application of the traditional view to God's action is more remarkable. Thanks to Tad Schmaltz for a discussion of this point.

[60] In this regard, as in many others, Newton's view has important affinities with Henry More's conception of God, and of course Newton may very well have been influenced by More himself. Compare the passage from Newton's (anonymous) "Account" with More's first letter to Descartes, sent in December of 1648:

> And, indeed, I judge that the fact that God is extended in his own way follows from the fact that he is omnipresent and intimately occupies the universal machine of the world and each of its parts. For how could he have impressed motion on matter, which he did once and which you think he does even now, unless he, as it were, immediately touches the matter of the universe, or at least did so once? This never could have happened unless he were everywhere and occupied every single place. Therefore, God is extended in his own way and spread out; and so God is an extended thing [*res extensa*]. (*Œuvres de Descartes, AT*, vol. V: 238–9)

> Hence Newton might be read as endorsing More's view that God cannot act on material objects unless God is spatially present to them, that is, unless they are not distant from God. In his response to More on 5 February 1649, Descartes obviously rejects this argument, noting, among other things, that it may presuppose the false view that nothing can exist, including even God, without being imaginable, a view that suggests, according to Descartes, that each thing must be extended – see *Œuvres de Descartes, AT*, vol. V: 270–1. For a discussion of this letter, and of More's relation to Descartes, see Garber, *Descartes's Metaphysical Physics*, 144ff. I follow Garber's translation above.

construal of "incredible" here involves the standard meaning, especially in the early modern period, of a phenomenon that is beyond belief.[61]

The implication of these elements for understanding Newton's over-arching view of action at a distance is clear: no material objects could possibly act at a distance on other objects, for then they would exhibit a miracle that, from Newton's point of view, even God does not exhibit.[62] Presumably, for a material object to exhibit miraculous behavior just is for that object to interact causally with other elements in the natural world in a fashion that violates the ordinary course of nature – expressed, perhaps, through physical laws – as a result of God's intervention. Hence for material objects to interact distantly, and thereby to interact miraculously, is not only for them to exhibit non-natural or non-law-like behavior, but for them *ipso facto* to exhibit God's intervention. But if God does not act at a distance on any object, it remains unclear how material objects could engage in such action by exhibiting God's intervention in the physical world. Such a miracle, therefore, seems to be ruled out by Newton's views.[63]

This point is connected, in turn, to Newton's own brief but significant speculation that an "immaterial agent" might serve as the physical basis of gravitational interactions. Since God is not distant from any object at any time, and since Newton obviously thinks that God might be the very "immaterial medium" underlying all gravitational interactions among mat-erial bodies, it seems clear that when Newton contemplates the idea that God might be the relevant mediating element he is not contemplating the idea that bodies act at a distance on one another.[64] Instead, God acts locally and directly on any object at any time. In an unpublished addition to the

---

[61] The *Oxford English Dictionary* gives "beyond belief" as the first definition of "incredible."

[62] See Koyré, *Newtonian Studies*, 16. But cf. Funkenstein, *Theology and the Scientific Imagination*, 96 for a different view.

[63] Of course, material objects might apparently change one another's states of motion at a distance if God were to do the causal work; but if God were acting locally on each object, then God's action would not constitute an "incredible miracle," according to Newton.

[64] As Newton writes to Bentley in February of 1693, continuing the passage quoted in the text above, "Gravity must be caused by an agent acting constantly according to certain laws; but whether this agent be material or immaterial, I have left to the consideration of my readers" (*Philosophical Writings*, 103). And David Gregory reports Newton's views as follows:

The plain truth is, that he believes God to be omnipresent in the literal sense; And that as we are sensible of Objects when their Images are brought home within the brain, so God must be sensible of every thing, being intimately present with every thing: for he supposes that as God is present in space where there is no body, he is present in space where a body is also present. But if this way of proposing this his notion be too bold, he thinks of doing it thus. *What Cause did the Ancients assign of Gravity* [?]. He believes that they reckoned God the cause of it, nothing els [sic], that is no body being the cause; since every body is heavy. (Hiscock, *David Gregory, Isaac Newton, and their Circle*, 30)

corollaries to propositions 4 to 9 of book III written in the early 1690s, Newton draws this connection explicitly:

For two planets separated from each other by a long empty [*vacui*] distance do not attract each other by any force of gravity or act on each other in any way except by the mediation of some active principle [*movente principio*] interceding between them by which the force is transmitted from one to the other. And therefore those ancients who rightly understood the mystical philosophy taught that a certain infinite spirit pervades all space & contains and vivifies the whole world [*spiritum quondam infinitum spatia omnia pervadere & mundum universum continere & vivificare*] …[65]

There can be no doubt, then, that Newton himself connected his denial of action at a distance with his conception of God's spatiotemporal ubiquity and corresponding potential role as a medium for all gravitational interactions.

In this case, then, we can understand a central aspect of Newton's natural philosophy only if we consider his conception of God. For only then can we decisively rule out the notion that bodies can exhibit action at a distance. Although Newton clearly rejects action at a distance – and indeed, in the strongest terms, on grounds of inconceivability – in his correspondence with Bentley in 1692/3, and although all of his statements in the *Principia* are perfectly consistent with this rejection of distant action, if we leave aside his understanding of God there is no decisive rejection of such action that was published within Newton's lifetime. However, if we consider Newton's published remarks quoted above, and take seriously his contention that an

---

Thus Newton may very well have contemplated the idea that God is directly responsible for gravitational interactions, and would therefore be the "immaterial" medium mentioned in the correspondence with Bentley – indeed, Bentley himself made this inference in his eighth Boyle Lecture (see *Isaac Newton's Papers and Letters*, 344). On Newton's attempt to revive various aspects of what he took to be ancient philosophical views, see Snobelen, "'The True Frame of Nature.'" Cf. also Westfall, *Force in Newton's Physics*, 399–400. Fatio reports that Newton "often" thought that "gravity had its foundation only in the arbitrary will of God" (Gagnebin, "De la cause de la Pesanteur," 117). In his notes on Leibniz's *Tentamen* Newton objects that Leibniz's mechanist presuppositions, discussed in the text above, prevent him from taking God to move the planets, so he clearly wants to leave that possibility open – *Correspondence*, vol. VI: 117. For his part Leibniz speculated toward the end of his life that Newton may in fact have thought of God as underwriting all gravitational interactions (see his last letter to Clarke, section 118; *Die philosophischen Schriften*, vol. VII: 418). See also Metzger, *Attraction universelle*, part 2: 75 n. 7.

[65] Unpublished manuscript in University Library Cambridge, Add. MS 3965.6, f.269; the original Latin is quoted in Casini, "Newton: The Classical Scholia," 38. I have slightly altered the translation of this passage given by Westfall, *Never at Rest*, 510–11. For discussion of these unpublished additions to book III, often called the "classical scholia" because of their many references to classical (ancient) textual sources, see McGuire and Rattansi, "Newton and the 'Pipes of Pan'"; for criticisms of their interpretation, see Casini's paper cited above.

analysis of the divine being is a proper part of natural philosophy, then DiSalle's interpretation may be undermined.[66]

But perhaps DiSalle's view can be rescued by the contention – found especially in Stein's work – that even Newton's conception of God should be regarded as fundamentally empirical in character. If Newton's principal understanding of God were empirical in some substantive sense, then the objection to DiSalle might be vitiated, for then the Newtonian view that even God was a local actor would also turn out to be a merely empirical claim, one subject to revision under certain circumstances.

Perhaps the most relevant aspect of Newton's conception of God is found in the General Scholium, in a passage quoted above: "It is agreed that the supreme God necessarily exists, and by the same necessity he is *always* and *everywhere*" (*Principia*, 942). In keeping with his general orientation toward Newton's work, Stein regards even this view as empirical.[67] He argues that Newton's overarching conception of God is founded either on perceptual experience or on revelation, where the latter is also regarded as a source of empirical knowledge.[68] And Newton does in fact insist that "to

---

[66] Despite the fact that DiSalle discusses *De Gravitatione* in some depth, and despite the fact that he (rightly, in my view) rejects the view that Newton defends what is now called "substantivalism," he then overlooks Newton's famous contention that God occupies all places within infinite space at all times, a contention found not only in unpublished works such as *De Gravitatione*, but even in the General Scholium to the *Principia*. See DiSalle, *Understanding Space–Time*, 26–7 on the structure of space, and 34–8 on the treatment of space in *De Gravitatione*. I discuss the question of Newton's adherence to a "substantivalist" conception of space in more depth in ch. 5. Stein tackles these aspects of Newton's views more directly.

[67] Stein buttresses his overall interpretation by contending that Newton never mentions any "*a priori* epistemological ground for any item of knowledge." Newton's extensive reading of Descartes's *Principia philosophiae* and *Meditations*, of course, would certainly have familiarized him with a classical rationalist conception of such an epistemological ground. However, this is far from decisive, for two primary reasons. First, despite his extensive attempts at undermining Cartesian views – of space, motion, etc. – Newton never addresses overarching epistemic issues concerning the basis of our knowledge of space, either in the Scholium or in *De Gravitatione*. His silence would appear to be neutral on the issue that Stein raises. Second, Newton also never addresses the question of whether his affection thesis, or his more general conception of space in *De Gravitatione*, is an empirical or *a priori* claim; he simply ignores this question. So this does not tell for or against Stein's view.

[68] Stein, "Newton's Metaphysics," 269–70. Stein writes:

besides knowledge of God "from the appearances of things," which "belongs to Natural Philosophy," Newton holds that there is knowledge of God through *revelation*. This, too, of course, would be through *experience*; and what is more important so far as concerns Newton's own efforts in the domain of "revealed" theology (efforts that occupied no small part of his whole intellectual career), the deliverances of revelation are, for Newton, accessible *only through historical documents* (Newton does not subscribe to any claim of immediate religious authority – nor does he claim access to revelation through personal inspiration), and therefore demand a very arduous historical-critical investigation of such documents. ("Newton's Metaphysics," 298 n. 19).

McGuire apparently agrees with Stein's general assessment that even this aspect of Newton's views should be seen as empirical – McGuire, *Tradition and Innovation*, 34–5.

treat of God from phenomena is certainly a part of natural philosophy"
(*Principia*, 943). Stein is presumably correct in maintaining that for Newton
our knowledge of God is not derived from "pure reason."[69] Stein's view
reflects Newton's consistent empiricist rhetoric, and his concomitant rejection of the view that we have an innate idea of God or indeed any *a priori*
source for our knowledge of the divine being.

However, Stein's interpretation may fail to reflect Newton's own orientation toward what we might call early modern epistemology, broadly
construed. Newton's occasional rhetoric aside (cf. *Principia*, 795–6), he in
fact has little to say about the sorts of epistemological question tackled by
Descartes or Locke; his focus lies instead on a distinct epistemic issue
generated by physics. The issue for Newton is not whether our views are
empirical in the sense that they originate from an *a posteriori* source, or
whether we have innate ideas, but whether our views are subject to procedures of refinement and revision. It is a hallmark of his work in natural
philosophy that we must always regard our conclusions regarding phenomena as subject to further refinement, or to fundamental revision, on the basis
of empirical evidence.[70] Precisely because this topic is so central to understanding Newton's achievement in the *Principia*, I can provide only the
briefest of sketches here; even a brief sketch, however, is sufficient to
engender doubts about Stein's interpretation.

From Newton's point of view, a proposition might be rendered more precise
in the light of new evidence in various ways. Consider just two: (1) the
exactness of a claim might be quantified – we might learn (e.g.) that a force
is proportional to $1/r^{2+n}$, rather than to $1/r^2$, or we might obtain a new
significant digit for a value;[71] and (2) we might revise a proposition to include
newly discovered exceptions. As for (1), for instance, when Newton argues in
the first edition of the *Principia* that the moon is kept in its orbit by a $1/r^2$ force,
he indicates that the data do not fit this conclusion exactly; rather, they
indicate that the force in question is 60¾ times closer to a $1/r^2$ than to a $1/r^3$
force. In the third edition he alters his calculation, indicating that it is 59¾
closer to $1/r^2$ than to $1/r^3$ (*Principia*, 802–3). As for (2), for instance, Newton
encapsulates his attitude in his fourth rule, which was added to the third

---

[69] Stein, "Newton's Metaphysics," 262.

[70] Perhaps more than any other contemporary commentator, George Smith has indicated the importance of understanding Newton's extraordinarily careful treatment of the relation between a proposition and the available evidence supporting it. See especially the illuminating discussion in his "The
Methodology of the *Principia*."

[71] In his Boyle Lectures of 1704, published as *A Demonstration of the being and attributes of God*, Samuel
Clarke explores this possibility (esp. 49).

edition of the text but which certainly governed his work in the early versions as well:

In experimental philosophy, propositions gathered from phenomena by induction should be considered either exactly or very nearly true notwithstanding any contrary hypotheses, until yet other phenomena make such propositions either more exact or liable to exception. (*Principia*, 796)[72]

Newton claimed that even the laws of motion were "deduced from the phenomena" (*Principia*, 943) and were therefore subject to revision or rejection. It seems reasonable, then, that Newton would regard refinement and revision of a proposition to be crucial elements in our epistemic attitude toward it.[73] This is his clear focus, rather than broad questions about the source of our knowledge.

Newton's own focus suggests a clear distinction between our fundamental knowledge of God on the one hand, and the rest of our knowledge in natural philosophy on the other. Newton's knowledge of God appears to be essentially fixed: it is presumably not subject to any procedures of refinement, revision, and rejection, for Newton holds what he acknowledges to be the traditional view that God is a necessary being, and there is certainly no suggestion in his work that we might revise this judgment on the basis of further evidence, concluding (say) that God was actually a contingent being, or that no necessary being existed (see *Principia*, 942). Indeed, it remains unclear what it would mean to undertake any revision of the judgment, short of simply rejecting it; and Newton never contemplates such a rejection during his long career. This presumably reflects the fact that for Newton, as for his principal interlocutors and critics, it is unclear how one could investigate natural phenomena in a way that would enable one to assess the existence of the necessary being. What would count as evidence that no such being existed after all? Given these points, the knowledge of God appears to be distinct in kind from the rest of our knowledge in natural philosophy. This is presumably more significant than the empirical source of our knowledge of God.[74]

---

[72] This rule is anticipated in an unsent draft of a letter to Cotes that Newton wrote in March of 1713 (*Philosophical Writings*, 119–22). See also the similar discussion in query 31, *Opticks*, 404, which first appeared in the 1706 edition.

[73] Newton's discussions of epistemology in other texts point in the same direction. For instance, he emphasizes in *De Gravitatione* that some of his discussions, such as the treatment of the creation of body, is merely probable while other parts, such as the analysis of space, appear to be more secure (*De Gravitatione*, 27). But he ignores the question of the source of our idea, and knowledge, of space. Thanks to Eric Schliesser for raising this point.

[74] There are further complexities here. For instance, DiSalle provides a compelling discussion of the fact that Newton transforms the age-old question of the earth's motion, and its placement in the solar system, into an empirical question concerning the center of mass of the solar system, noting that

This suggestion is perfectly consistent with the aspects of Newton's conception of God that Stein emphasizes. That is, emphasizing that Newton's conception is not subject to refinement and revision is consistent with Newton's view that one can undertake an investigation of God by studying the "phenomena." Our view that God is a necessary being – and that God must therefore necessarily be spatiotemporally ubiquitous – is not subject to revision, but we can make evidentiary arguments regarding the existence of such a being. For instance, in the General Scholium of 1713 Newton makes the familiar teleological argument that the astonishing complexity among various flora and fauna could only have arisen through the intervention of a necessary being – hence he "treats" of God from the phenomena.[75] Yet this does not entail that his view that God is a necessary being is itself open to revision, even if this argument fails; and it would presumably not be open in that way even if the phenomena were radically different and lacked the relevant complexity. Consider another, slightly different, example: in the correspondence with Bentley, Newton makes another teleological argument, one regarding the "system of the world." He contends in his second letter to Bentley at the end of 1692 that "the motions which the planets now have could not spring from any natural cause alone, but were impressed by an intelligent agent" (*Philosophical Writings*, 95). Here, too, one encounters an argument concerning God founded on an analysis of the phenomena. Yet the view that God is a necessary being, and therefore gave rise to all contingent beings in the world, would presumably not be in jeopardy if the solar system were to have an arbitrarily different planetary structure. For Newton never indicates how natural phenomena such as the structure of the solar system could falsify his conception of the necessary being.

If we place Newton's understanding of God at the center of his metaphysical system we obtain a picture of a nuanced and complex conception of

---

nothing at all is at that center of mass, but the sun is very close to it, and much closer than any of the planets, including the earth. And there is a straightforward sense in which this is an empirical question for Newton. However, Newton also approaches what we might call the structure of the world system – more precisely, the spatial placement of the planets *vis-à-vis* one another and the sun – in a rather different way, concluding that only God could possibly have established the structure of the solar system, and he does so in the first edition of the *Principia* in 1687. In the first edition, Newton argued that God "placed" the planets in their current orbits according to their degrees of density. This was in corollary five to proposition eight of book III – see Cohen, "The Review of the First Edition of Newton's *Principia*," 340, for discussion. It seems that this represents an empirically based argument concerning God's causal influence on the structure of the solar system. This kind of argument may represent a sort of hybrid of what I have called physics and metaphysics in the text above.

[75] See *Principia*, 942; Newton makes a parallel argument in query 28, *Opticks*, 370.

metaphysics in relation to physical theory.[76] In the broadest terms this conception can be described as follows: the metaphysical aspects of natural philosophy are bifurcated into what we might call divine and mundane metaphysics. Divine metaphysics, as we have seen, represents a fundamental conception of God's nature and relation to the natural world that is not subject to revision; hence it might be understood to represent a basic framework for all of Newton's thinking about the physical world, one that is never questioned as he progresses through numerous empirical and mathematical investigations. Mundane metaphysics occurs within the basic framework centered on the divine; it is subject to precisely the sorts of revision and refinement that characterize all of Newton's other work. Mundane metaphysics concerns metaphysical issues not directly focused on the divine: the nature of motion, the existence of various types of forces in nature, the types of causation involved in natural change, and so on. This allows me to articulate my limited acceptance of the Stein–DiSalle interpretation: Newton does in fact transform the issues of mundane metaphysics into empirical questions in just the sense that our answers to these questions are always subject to refinement, revision, and rejection on the basis of empirical evidence. But this research into nature occurs within a fundamental framework that is fixed independently of any empirical findings we make.

An example might illuminate this interpretation – consider again Newton's discussion of action at a distance, on the one hand, and of causation on the other. As we have seen, Newton takes all entities, including even God, to engage only in local action – this general conception represents an element in his fixed framework centered on divine metaphysics. Yet the question of whether all causation must be mechanical is transformed from the *a priori* issue articulated by the mechanists into a thoroughly empirical question, one answered in fact through Newton's theory in the *Principia*. Hence this represents an issue within mundane metaphysics. Whereas empirical scientific research presumably could never discover action at a distance in the physical world, from Newton's point of view, it certainly could discover non-mechanical causation.[77] The development of

---

[76] For an extraordinary discussion of the relation between Newton's theory of gravity on the one hand and various metaphysical and theological issues on the other, see Metzger, *Attraction universelle*. She provides an insightful analysis not only of Newton's own views, but of the views defended by many of his prominent eighteenth-century commentators and defenders, including Richard Bentley and Samuel Clarke, discussing what she calls the "theological variations on the theme of universal attraction" (*Attraction universelle*, part 1: 48).

[77] I support this possibly controversial contention in ch. 6.

physical theory must be consistent with the metaphysical framework that
Newton embraces, but that framework itself does not dictate the questions
or the answers that physical theory will provide in the arena of remaining
metaphysical issues. That represents a revolution in conceptions of the
relation between physics and metaphysics in the late seventeenth century.

But this point raises a problem for my interpretation that the Stein–
DiSalle reading lacks. If Newton's metaphysical framework characterizes all
action, even divine action, as local in character, how does his view differ
from the mechanist requirement that there be no distant action, and that
therefore – from their point of view – all action is by contact on surfaces of
bodies? It differs in two fundamental ways, one substantive and the other
methodological.

First, the substantive point: the mechanists conflate local action with
surface action; Newton teases these apart, accepting that a force such as
gravity is non-mechanical (it does not operate on surfaces) but nonetheless
purely local, perhaps involving an ether that flows through material objects,
interacting not with their surfaces but with their masses. Second, Newton's
methodology remains distinct from that of the mechanists; he never
employs his rejection of distant action to underwrite a particular hypothesis
or explanation regarding (e.g.) the cause of gravity. This differs fundamentally from the employment of mechanist principles in, say, the *Tentamen*
(1689),[78] where Leibniz attempts to show that the planetary orbits are the
result of the motions of swirling vortices by citing the principle that no body
undergoes natural change except through surface action.[79] Thus although
Newton never accepts distant action, he avoids "feigning" or endorsing any
hypothesis – such as his own ether theory, outlined in query 21 to the
Opticks – concerning the cause of gravity.[80]

This second (methodological) point, in turn, helps to highlight the
complexity of Newton's understanding of the relation between metaphysics and physical theory, a complexity demanded in fact by Newton's own

---

[78] *Die mathematische Schriften*, vol. VI.

[79] Leibniz was certainly not alone in this regard: in *Occult Powers and Hypotheses*, Clarke convincingly
argues that the Cartesians introduced hypotheses – he even says that they endorse *hypotheses fingo* –
precisely because of their commitment to mechanical explanations, and possibly in part because of
their similar endorsement of the view that matter's essence is extension.

[80] Moreover, why can Newton in book II and in the General Scholium (and elsewhere, for that matter)
discuss the vortex theory of the Cartesians as a competitor to his own account of gravity when the
former is a physical treatment of force and the latter a merely mathematical one? That is to say, on
Newton's own view, at least prima facie, these two theories have distinct aims and would not appear
to be competitors to one another. My answer is this: since the vortex theory most definitely involves
hypotheses, it is clear that Newton's theory, which includes a meta-level perspective on the
introduction of hypotheses into experimental philosophy, competes with it.

theory of gravity. As it turns out, of course, Newton's theory in the *Principia* is perfectly consistent with the view – which some philosophers, such as Boscovich and Kant, endorsed in the later eighteenth century – that material bodies such as the planets act on one another directly and independently of any medium between them.[81] That might even represent the most natural interpretation of his gravitational theory. And more generally, his theory is also perfectly consistent with the view that God exists beyond the bounds of the physical world, or with a deeply atheist conception of the origins of the universe. In that regard the divine metaphysical framework I have attributed to Newton is fundamentally distinct from a Cartesian conception of the metaphysical foundations of physics: Newton's physical theory does not presuppose the basic elements of the metaphysical framework in which it is articulated, at least in the sense that there are interpretations of the theory that entail the falsity of one or more of those elements. Rather, from Newton's perspective, the metaphysical framework dictates his preferred interpretation of his theory, and settles issues – such as God's action within the physical world – that are unanswered by the physical theory. The only constraint that the framework might impose is to rule out any physical theory that demands its falsity: if a new physical theory were to require distant action explicitly, or to deny God's role as the creator of the universe, then it is difficult to see how Newton could endorse it.[82]

This last point, in turn, suggests that we might characterize Newton's understanding of the relation between physics and metaphysics in more detail by contrasting it with the basic elements in Descartes's system. We know from Daniel Garber's work that Descartes develops what we can usefully characterize as a "metaphysical physics," especially in his *Principia Philosophiae*, a text well known to Newton. This might mean, at least in part, that Descartes takes physics to have a metaphysical foundation in the sense that metaphysics is logically prior to physics. The Stein–DiSalle reading might then be said to characterize Newton as inverting the elements in Descartes's view, so that physical theory is logically prior to

---

[81] As I discuss in greater detail in ch. 6, Newton seems committed – in his argument for universal gravity in book III of the *Principia* – to a direct momentum exchange between the planetary bodies, but not to the view that these bodies exchange momentum by acting directly on one another across empty space, independently of any medium between them. For the momentum exchange is consistent with the possibility that some medium, perhaps even a divine one, somehow transfers causal influences among the bodies. Many thanks to Michael Friedman for clarifying these issues for me – see especially his treatment of these matters in "Newton and Kant."

[82] See ch. 6 for further discussion of the points in this paragraph.

metaphysics.[83] Hence Newton would have a physical metaphysics. But as we have seen this reading misses the fact that Newton's conception of God's fundamental relation to the natural world is in fact not logically posterior to physical theory, for it is immune to revision regardless of any developments within physics. So if Newton presents a physical metaphysics to rival Descartes's metaphysical physics, he does so within crucial parameters. Only his mundane metaphysics is physical in the sense of being logically posterior to physics.[84]

One reason to think of the conception of God found in Newton's writings as constituting a metaphysical framework is that Newton provides God with a metaphysical, but not an epistemic, primacy, a fact that underscores his difference from Descartes. For Descartes, God is not only metaphysically primary, but epistemically primary: in book II of *Principia Philosophiae*, we derive the first two laws of nature from God's property of immutability (perhaps with further premises), a property known independently of the laws in particular and of physical theory more generally.[85] This derivation not only provides us with the content of the laws of nature, but also with an answer to a more fundamental question, namely, why are the laws of nature the way they are? The answer mimics our derivation of the laws: since the laws reflect God's immutability, they must have the content that they do. But for Newton no such derivation is considered, and although our knowledge of God is explicitly mentioned in the *Principia* – for instance, in the General Scholium – it certainly plays no role in the derivation of any proposition within that text, including the laws of motion. Newton would have rejected the very idea that we can explain the content of

---

[83] This allows me to clarify an issue broached in ch. 1. Newton obviously does not locate himself within the seventeenth-century metaphysical tradition as Descartes locates himself, so if this interpretation is to be persuasive, we have to acknowledge that Newton does not simply differ with Descartes on the relation between physics and metaphysics; he also rejects Descartes's understanding of the relata. Even if we restrict ourselves to the *Principles* – leaving aside such texts as the *Meditations*, which Newton had in his library, but which he did not discuss in any depth – we find that Descartes tackles numerous metaphysical issues on which Newton remains silent. Since Newton attempts systematically to refute various Cartesian conceptions in *De Gravitatione*, we might reasonably employ it as a guide here: it expresses his agnosticism on many of the issues tackled in the *Principia Philosophiae*. In *De Gravitatione*, Newton refers repeatedly to articles and passages within parts II and III of the *Principia Philosophiae*, but does not refer to part I. Instead, he focuses almost exclusively on the questions about body, space and motion broached in parts II and III; so he sees the metaphysics of natural philosophy as somewhat restricted.

[84] For physics to be logically prior to Newton's mundane metaphysics is, in part, for the latter to be subject to refinement, revision, and rejection on the basis of developments in physics. Newton's discovery of non-mechanical causation is a clear example.

[85] For a helpful discussion of the derivation, including the derivation of the third law from our innate concept of body, see Nelson, "Micro-Chaos and Idealization in Cartesian Physics."

the laws of nature through some more fundamental divine principle or attribute; we discover the laws of nature through empirical research and therefore lack any explanation of their content. The same is true for other facts: gravity could have been a $1/r^{2+n}$ force; we lack any explanation of why it deviates so little from the inverse square.[86] So God's relation to the "system of the world" forms a framework within which physical research takes place, but the elements of that relation play no role in guiding that research. Newton obviously takes God to have created the universe and to have decreed its laws, but these facts do not steer our research.

Newton's view that God is a necessary being clearly entails that there are no brute facts: all the facts that characterize the physical world, including the laws of nature, are contingencies that follow from God's will, and God alone exists necessarily. Although we know that the laws of nature are not brute facts, however, this knowledge does not enable us to determine why God chose the laws that govern our world.[87] Since our knowledge that there are no brute facts follows from our unrevisable knowledge that God is a necessary being, it (too) is presumably unrevisable. All contingencies concerning the state of our world flow from the one necessary being. Yet for all the facts concerning the phenomena – including other facts about the laws and their content – our knowledge is subject to refinement, revision, and rejection.

Of course, thus far I have only sketched my interpretation of Newton's divine and mundane metaphysics, providing only a minimal picture of the way in which it may account for the shortcomings of the Stein–DiSalle interpretation. The claims I have made thus far require much more detailed argumentative support, and raise numerous questions of their own. Each of the next four chapters provides some of that support, and answers some of those questions. Roughly speaking, chs. 3 and 4 focus on central elements of Newton's mundane metaphysics, and ch. 5 highlights elements of his divine metaphysics. In ch. 6 I then outline how these two aspects of his thought intersect.

---

[86] By 1903, painstaking tests had indicated that gravity deviates from being a $1/r^2$ force less than two parts in 10^8. Thanks to George Smith for a memorable lecture on these, and related, empirical issues.

[87] If Newton endorses the view of divine will that Clarke articulates in his correspondence with Leibniz, then it may also be true that what God wills – including the laws of nature – is itself a brute fact. That is, God's willing would be a brute fact if it were not governed by and explicable via the principle of sufficient reason, or some other sufficiently general metaphysical principle. Leibniz would obviously reject this view out of hand. See, e.g. C 2: 1, L 3: 7, C 3: 2, etc. Thanks to Tad Schmaltz for raising this point.

3

Do forces exist? contesting the mechanical philosophy, I

> For many things lead me to have a suspicion that all phenomena may depend on certain forces by which the particles of bodies, by causes not yet known, either are impelled toward one another and cohere in regular figures, or are repelled from one another and recede. Since these forces are unknown, philosophers have hitherto made trial of nature in vain. But I hope that the principles set down here will shed some light on either this mode of philosophizing or some truer one.
>
> – Preface to first edition of the *Principia*

It is widely recognized that Newton's *Principia* helped to bring the vocabulary of "force" into what was then called natural philosophy and, later, into what became mathematical physics. Newton declares that he composed the *Principia* precisely to determine the forces of nature and, of course, the highlight of the *Principia* is the so-called derivation, in its third and final book, of the law governing the force of gravity.[1] Yet Newton's various pronouncements regarding his "mathematical" treatment of force – mentioned already in ch. 2 – often muddy the waters. In fact there is no consensus in the scholarly literature regarding even the most basic question concerning the mathematical treatment: does Newton think that forces exist? Does he think in particular that the force of gravity exists? Consider, for instance, the stances of the two most influential interpreters of Newton among historians of science in the twentieth century: on the one hand, I. B. Cohen contends that Newton never even addressed the question of the "existence of forces"; on the other hand, in his magisterial *Force in Newton's Physics*, Richard Westfall declares that Newton not only addressed the question, but fundamentally rejected the prevailing mechanical philosophy

---

[1] In the author's preface to the first edition, Newton writes: "For the basic problem of philosophy seems to be to discover the forces of nature from the phenomena of motions and then to demonstrate the other phenomena from these forces" (*Principia*, 382).

of his day by insisting that forces must be considered part of our fundamental ontology.[2]

To cope with this interpretive quandary I intend to follow the familiar Kantian dictum of splitting the difference, accounting for this ground-level disagreement by arguing that Cohen and Westfall each capture, and each miss, something crucial about Newton's attitude toward forces.[3] In particular, although I take Newton to have fundamentally rejected the mechanical philosophy, that is not because, for instance, he considered some mechanist proposition and then denied its truth; instead, I take him to eschew something more fundamental, namely the way in which the mechanists ask questions about ontology and, even more importantly, the way in which they demand that questions about ontology be answered. If we follow this lead, I think, we will see how both Cohen and Westfall capture important aspects of Newton's views, but also leave something unexplained.

Because of the *Principia*'s focus on gravity, there are actually two interrelated, but separable, issues about force. One can effectively separate the two issues by distinguishing between two versions of the mechanical philosophy embraced by some of Newton's principal critics.[4] Hence the goal here is not to categorize or characterize seventeenth-century mechanists

---

[2] See Cohen, "The Review of the First Edition of Newton's *Principia*," 347 and Westfall, *Force in Newton's Physics*, 377.

[3] Kant endorses this approach not only in the *Critique of Pure Reason*, but also in the so-called precritical works, including even his very first essay of 1747, on living forces – see *Kants gesammelte Schriften*, vol. I: 32.

[4] There are other versions of the mechanical philosophy that I do not discuss here; I discuss further nuances in understanding "mechanism" in ch. 4 below. McGuire provides a helpful delineation of the various mechanist views available in the seventeenth century:

> It is obvious that the term "mechanical" meant many different things to various thinkers of the seventeenth century: nature is governed by immutable geometrical laws; contact action is the only mode of change; first principles are to be integrated with experimental investigations; regularities are to be explained in mathematical form; that all phenomena arise from matter *in* motion, or matter *and* motion; that compound bodies are composed of vortices (Descartes), centers of force (Leibniz), or tiny bits of matter conceived as atoms or corpuscles; that changes in phenomena result from the *way* in which internal particles alter their configurations; that the "new science" conceives nature dynamically in terms of motion, rather than statically in terms solely of the size and shape of internal particles; that occult qualities are to be banished from explanations which must be based on sensory experience in terms of clear and distinct ideas; or that nature is to be conceived in analogy to the operations of mechanical activities. All of these characteristics and others – some of which were emphasized more by one thinker than by others – characterize what Boyle called the "Mechanical Hypothesis or Philosophy," *Works*, I, 230, n. 5. Thus, while they all agreed that contact action was a necessary condition for a mechanical explanation, there was no settled agreement as to sufficient conditions. (McGuire, "Boyle's Conception of Nature," 523 n. 2)

> I concur with McGuire that contact action was a "necessary condition" for mechanical explanation – hence both the strict and the lenient mechanist (described in the body of the text above) require contact action. Thanks to Karen Detlefsen for discussion of this point.

exhaustively, but to indicate features of their views that are salient for understanding Newton. First, there is a general issue concerning force: adherents of what I will call strict mechanism reject the very idea of forces, arguing that all natural change occurs through impacts among material bodies characterized exclusively by size, shape, and motion (and perhaps solidity, *à la* Locke).[5] For the strict mechanist any talk of forces must be merely instrumental or provisional, which may place all of Newton's dynamics in philosophical jeopardy. Second, there is a special issue concerning the force of gravity: adherents of what I will call loose mechanism may admit the intelligibility of forces, insisting only that no natural change among material bodies occurs through action at a distance. Thus the mature Leibniz places forces at the center of his metaphysical system, but rejects what he takes to be the Newtonian reliance on action at a distance within gravitational theory.[6] Both strict and loose mechanists reject action at a distance, then, but only the former jettison all forces. For the strict mechanist, Newton's treatment of gravity is troubling because it represents a particular instance of a general problem regarding forces; for the loose mechanist, gravity in particular is problematic because it seems to involve genuine attraction or distant action between material objects. Any force involving impact alone is perfectly acceptable.

The mechanist criticisms of the *Principia* are legion and involve numerous philosophical issues. One especially subtle and deep criticism, raised by

---

[5] Attributing this conception of the mechanical philosophy to any given canonical figure, of course, will involve a certain degree of contestable interpretation. But it seems reasonable to think of Boyle and Locke along these lines, and potentially the mature Huygens. See, for instance, Boyle's characterization of a mechanical view of nature in *A Free Enquiry into the Vulgarly Received Notion of Nature* (*Works*, vol. IV: 372). I discuss Locke's views briefly below. For Huygens's mature view, see the opening paragraph of his *Discours sur la cause de la Pesanteur* (1690), in which he articulates a mechanist constraint on any intelligible explication of gravity – Huygens, *Discours*, in *Œuvres complètes*, vol. XXI: 129 (the latter is the pagination of the original 1690 edition, provided marginally). For discussions of Huygens's perspective on forces, see Westfall, *Force in Newton's Physics*, 161–3 and De Gandt, *Force and Geometry*, 139. Descartes, of course, is a special case because of his formulation of the third law of nature (*Principia Philosophiae*, VIII-1: 65). I discuss Descartes's views in greater depth below.

[6] In the context of criticizing Newton's gravitational theory along familiar lines, Leibniz provides a clear expression of what I have called loose mechanism in his "Anti-barbarus physicus," noting that all "true" corporeal forces must involve impetus – Leibniz, *Philosophical Essays*, 313, 319. Because of the complexity of Leibniz's thought, rendered even more complex by the changes in his views over time, I present him in what follows primarily as a critic of Newtonian natural philosophy, a role he adopted in numerous public venues late in life. Therefore I bracket Leibniz's nuanced conception of forces and his equally nuanced conception of the limits of the mechanical philosophy, emphasizing his adoption of a loose mechanist perspective in his own account of the motions of the planetary bodies in the *Tentamen* and in his many criticisms of Newton. Since Leibniz wrote the *Tentamen* (1689) after reading the first edition of the *Principia*, it can be interpreted in the context of his dissatisfactions with Newton's theory. Thanks to Karen Detlefsen and Michael Friedman for discussion of these issues.

Leibniz among others, highlights the question of whether forces such as gravity exist. It does so by suggesting that Newton's treatment of gravity saddles him with a dilemma. If Newton contends that gravity exists, he must admit that material bodies act on one another at a distance, thereby violating a crucial norm of the mechanical philosophy (in all its guises). However, if Newton seeks to avoid action at a distance, as all of his contemporaries do, he must admit that gravity does not exist, and that he has therefore failed to discover the cause of the previously disparate celestial and terrestrial phenomena unified by his gravitational theory. In what follows I argue that Newton's general defense of forces against the strictures of strict mechanism enables him to evade this particular dilemma involving the force of gravity by rejecting the presuppositions underlying it.

NEWTON'S DILEMMA: ACTION AT A DISTANCE

Newton's mechanist interlocutors were well aware that the force of gravity raises special issues that other forces – such as those involving collisions between material bodies – do not, and they harnessed this fact in order to saddle Newton with a dilemma. In particular, Newton's theory treats gravity as if it involves action between extremely distant bodies, such as Jupiter and the sun. If Newton contends that gravity exists, they argued, he must be claiming that material bodies act on one another across empty space.[7] However, if he seeks to avoid distant action, he must admit that the force of gravity does not exist, and that he has therefore failed to discover the cause of the phenomena his theory associates with that force, such as the planetary orbits.[8] This is precisely the dilemma that Leibniz famously attempted to foist upon Newton.

---

[7] Part of the background here, of course, involves Newton's well-known failure to characterize (what Newton's critics would call) the mechanism underlying gravity. There is obviously much to be said about what counts as a mechanism, whether there are non-causal mechanisms, what renders an explanation, or a phenomenon, "mechanical," and so on. For a classic account of such issues, see Nagel, *The Structure of Science*, 153–202. We will see in what follows that by a "mechanism," Newton means a causal process that involves only surface action; the point is germane to his understanding of what kind of mechanism, if any, could account for gravity.

[8] See the classic accounts in Hesse, *Forces and Fields*, and Westfall, *Force in Newton's Physics*; see also McMullin, *Newton on Matter and Activity*, "The Explanation of Distant Action," and "The Origins of the Field Concept in Physics," 26. Putnam – in *Reason, Truth, and History* (200) – writes:

For example, if I make the inference from Newton's description of the solar system to the statement that "it is the gravitational attraction of the moon that causes the tides" then I am employing my informal knowledge that there is a conceptual link between statements about forces and statements of the form *A caused B*. The word "cause" does not even appear in Newton's description of the solar

Since Newton himself insists that he presents a merely "mathematical" treatment of force in book I of the *Principia*, it may seem natural to conclude that he aims to avoid action at a distance by denying that gravity actually exists, construing it perhaps as a mere calculating device.[9] Many prominent eighteenth-century Newtonians read the *Principia* in just this way, sometimes placing Newton within the "Galilean" tradition of investigating nature mathematically while bracketing causal issues.[10] This reading is bolstered, in turn, by Newton's famous proclamation to Richard Bentley that action at a distance is unintelligible (mentioned already in ch. 2). As we have seen, he wrote to Bentley in 1693, six years after the *Principia* first appeared:[11]

That gravity should be innate, inherent, and essential to matter, so that one body may act upon another at a distance through a vacuum, without the mediation of anything else, by and through which their action and force may be conveyed from one to another, is to me so great an absurdity that I believe no man who has in philosophical matters a competent faculty of thinking can ever fall into it. (*Correspondence*, vol. III: 253–4)

In forcefully rejecting the very idea of action at a distance, Newton appears to accept one horn of the dilemma. And it might seem reasonable to infer that this rejection entails Newton not conceiving of gravity as a genuinely existing force.

However, elsewhere Newton appears to avoid accepting this horn of the dilemma. As is well known, he acknowledges in the General Scholium that

system and of the tides; but I know that the gravitational force that *A* exerts on *B* can be described as caused by (the mass of) *A* simply by virtue of understanding Newton's theory."

In this passage, Putnam expresses a reasonable view of Newton's theory, and a reasonably common one.

[9] For discussions of this point, see Gabbey, "Force and Inertia in the Seventeenth Century," 239; Densmore, "Cause and Hypothesis: Newton's Speculation about the Cause of Universal Gravitation," 94–111; McMullin, "The Impact of Newton's *Principia* on the Philosophy of Science," 281; and Smith, "Comment on Ernan McMullin's 'The Impact of Newton's *Principia* on the Philosophy of Science,'" 327–38. Costabel comments that Newton thought of gravity as "mathematical" rather than as "real" in "Newton's and Leibniz's Dynamics," 111.

[10] The idea might be that Galileo's method departs from Scholastic tradition by being mathematical and by avoiding the investigation of causes, for instance the cause of free fall. For discussion of the evidence that Galileo employs a method that is mathematical and causally agnostic, see Douglas Jesseph's introduction to Berkeley, *De Motu*, 10–12ff.

[11] This chapter presupposes ch. 2 in the sense that it does not revisit the question of whether Newton accepted action at a distance; I take it as given here that he did not, citing only the Bentley letter as support. This is treated in greater depth in ch. 6.

he has failed to discover the "cause," or the physical basis, of gravity.[12] He then adds a rather striking contention, which bears quoting at length because it is nestled in between other famous pronouncements:

Thus far I have explained the phenomena of the heavens and of our sea by the force of gravity but I have not yet assigned a cause to gravity [*Hactenus phænomena cælorum & maris nostri per vim gravitatis exposui, sed causam gravitates nondum assignavi*]. Indeed, this force arises from some cause that penetrates as far as the centers of the sun and the planets without any diminution of its power to act, and that acts not in proportion to the quantity of the *surfaces* of the particles on which it acts (as mechanical causes are wont to do) but in proportion to the quantity of *solid* matter, and whose action is extended everywhere to immense distances, always decreasing as the squares of the distances . . . I have not as yet been able to deduce from phenomena the reason for these properties of gravity, and I do not feign hypotheses [*hypotheses non fingo*] . . . And it is enough that gravity really exists [*Et satis est quod gravitas revera existat*], acts according to the laws that we have set forth, and suffices for all the motions of the heavenly bodies and of our sea [*& ad corporum caelestium & maris nostri motus omnes sufficiat*].[13]

If Newton intended to treat gravity as purely mathematical, as a mere calculating device, we would not expect him to contend that it "really exists," a contention repeated in drafts related to the first edition of the *Principia*. In one such draft, an intended *Conclusio* for the first edition,

[12] Newton consistently calls the physical basis of gravity its "cause" in the *Principia*, the *Opticks*, and in unpublished manuscripts. See e.g. *Principia*, 382, 943; the first page of the advertisement to the reader in the third edition of the *Opticks* (1721) and *ibid.*, 351, 375, 377; and the drafts of the *Principia* in Hall and Hall, *Unpublished Scientific Papers of Isaac Newton*, 340–1.

[13] See *Principia Mathematica*, vol. II: 764. In *Principia*, Cohen and Whitman translate Newton's phrase "*& ad corporum caelestium & maris nostri motus omnes sufficiat*" as "and is sufficient to explain all the motions of the heavenly bodies and of our sea" (943), but the verb *explain* is missing from the original Latin. Motte (1729) in *Sir Isaac Newton's Mathematical Principles* has: "and abundantly serves to account for all the motions of the celestial bodies, and of our sea." Cohen's translation in this case may reflect his interpretation of the General Scholium, and of Newton's view of force more generally – for instance, in one essay, he writes:

But Newton, like Laplace and like the succeeding generations of mathematical physicists, did not become derailed by preliminary philosophical discussions of basic concepts. Rather, as we have seen, he explored the mathematical properties of systems of forces and particles or bodies without entering into discussions of the nature of forces, the modes of operation of forces, or the existence of forces. In the end, he produced a consistent and coherent system of mathematical physics that, as he was later to say in the General Scholium, serves fully to explain the motion of our seas and of the heavenly bodies. The same system also explained many aspects of terrestrial physical phenomena. As Newton said, in the General Scholium (1713), "Satis est," "That is enough!" (I. Bernard Cohen, "The Review of the First Edition of Newton's *Principia*," 347)

But it seems to me that: (1) "*satis est*" refers to the fact that gravity really exists, for the relevant sentence begins: "*Et satis est quod gravitas revera existat*"; and (2) Newton addresses what Cohen calls "the existence of forces" in this very sentence. Cf. the discussion in Harper and Smith, "Newton's New Way of Inquiry," 141–2.

Newton discusses what other forces – in addition to gravity – might be discovered through empirical investigation, using the same language as in the published text:

> I have briefly explained this, not in order to make a rash assertion that there are attractive and repulsive forces in bodies, but so that I can give an opportunity to imagine further experiments by which it can be ascertained more certainly whether they exist or not. For if it shall be settled that they are true [forces] it will remain for us to investigate their causes and properties diligently, as the true principles from which, according to geometrical reasoning, all the more secret motions of the least particles are no less brought into being than are the motions of greater bodies which we saw derived from the laws of gravity above. (Hall and Hall, *Unpublished Scientific Papers of Isaac Newton*, 327–8)

The insightful editor of the *Principia*'s second edition of 1713, Roger Cotes, a frequent correspondent of Newton's, highlights this aspect of Newton's view explicitly in his influential preface.[14]

Newton does not dodge the implication that if the force of gravity exists, it bears causal relations, for he concludes the Scholium to proposition 5 of book III with the dramatic pronouncement: "And therefore that force by which the moon is kept in its orbit is the very one that we generally call gravity" (*Principia*, 805). So although he is agnostic in the *Principia* on the underlying *cause* of gravity – by which he means its underlying physical basis, say in some medium between bodies – his agnosticism does not hinder him from claiming that gravity prevents the moon from following the inertial trajectory along the tangent to its orbit. He therefore does not appear to take the *Principia* to be neutral regarding the causes of certain motions, including the motions that constitute the lunar orbit. As some have argued, discovering such causes might be one of the text's primary goals.[15]

The interpretive difficulty we face, therefore, is rather stark. Newton appears to claim both that gravity exists – which means that it causes various natural phenomena – and that action at a distance must be rejected within natural philosophy. But given his agnosticism concerning the physical basis of gravity, how can he contend that it is causally efficacious without invoking action at a distance? How can it be specified as the cause of anything without characterizing its underlying physical basis, or what Newton sometimes calls its "physical seat"? If Newton is committed to finding the

---

[14] In his explication of Newton's conception of the force of gravity, Cotes writes of gravity that it "*revera existere*" – *Principia Mathematica*, vol. I: 27.

[15] See Stein, "On Philosophy and Natural Philosophy in the Seventeenth Century," 195.

"physical seat" of gravity, how can he proclaim in the General Scholium that gravity "really exists"? Given what Newton knew, even at the time of the third edition of the *Principia* in 1726, an ether might eventually have been discovered as the medium in question, in which case one would think that the proper conclusion would be that the ether – but nothing above and beyond it called universal gravity – is what "really exists."[16] One would presume that if the "physical seat" had yet to be discovered, referring to the force of universal gravity at all would have to be merely provisional, proper only for the purposes of calculation. And yet Newton's statements do not seem to be provisional.

This chapter attempts to answer this set of questions.[17] I argue that the solution lies in reinterpreting Newton's distinction in the *Principia* between what he calls the "mathematical" and the "physical" treatment of force.[18] As we will see, it may be natural to assume that the merely "mathematical" treatment of force brackets questions concerning causation, leaving them for the future "physical" treatment. One reason why that interpretation seems compelling is that there is a relevant historical precedent in astronomy – according to a prominent tradition, Copernican astronomy is a complex mathematical theory that eschews any causal account of the heavenly orbits. Hence it was said to "save" – but not to explicate the cause of – the phenomena.[19] However, I will suggest that Newton's mathematical treatment is intended to identify an existing force, a genuine cause of motion, and not merely to employ a calculating device. What the mathematical treatment of force leaves for future research is not the discovery of causes, in my view, but rather the discovery of the complete physical characterization of the force it has identified. This interpretation may indicate how Newton is able to avoid the dilemma with which Leibniz confronts him, but it will take the rest of the chapter to spell this out.

---

[16] I refer here to universal gravity to indicate that even if an ether were discovered, presumably one could use "gravity" to refer to the weight of bodies on earth, as the term was often used prior to the *Principia*. Thanks to George Smith for pointing this out to me.

[17] A caveat is in order here. In this chapter I do not discuss Newton's "derivation" of the law of universal gravitation in book III of the *Principia*, the validity of that derivation, or general aspects of the methodology he employs in his text (although cf. ch. 6 for a brief discussion of one aspect of the derivation). The literature on these topics is obviously voluminous, but see especially Stein, "Newtonian Space–Time," and Smith, "The Methodology of the *Principia*."

[18] On Newton's treatment of force before the *Principia*, especially in the so-called *De Motu* of 1684, see the outstanding discussion in De Gandt, *Force and Geometry in Newton's "Principia,"* 25ff., 48.

[19] The *locus classicus* for discussions of this view is Duhem, *Sozein ta Phainomena. Essai sur la notion de théorie physique de Platon à Galilée*. Cf. also the discussion in McMullin, "The Explanation of Distant Action," 290ff., and in "The Origins of the Field Concept in Physics," 22.

THE MATHEMATICAL TREATMENT OF FORCE

What is a force?[20] It is a cause of a change in a body's state of motion. Although force is a technical term for Newton – one that can be understood only through his laws of motion, which explicate force generally, and through his definitions, which define such particular items as centripetal force – we do not require a definition or technical understanding of causation to understand the *Principia*. If we leave aside the distinction between mathematical and physical treatments of force, which I discuss below, Newton deals with technical terms and concepts in the *Principia* in two distinct ways. In the definitions he introduces technical terms such as centripetal force, quantity of matter (mass), quantity of motion (momentum), and so on, that would have been unfamiliar to his readers (*Principia*, 403–8). And in the Scholium, which follows the definitions, Newton explicitly notes that space, time, place, and motion are understood by everyone, and therefore require no definition – but following Descartes's *Principia Philosophiae*, he adds that the "common" (*vulgare*) conception of these notions must be rigorously distinguished from the "mathematical" (or "absolute") conception of them (*Principia*, 408–9). Since Newton never defines "cause," or various related terms such as "action," and does not distinguish an ordinary from a more precise conception of causation, he apparently thinks that our ordinary conception is adequate for interpreting the *Principia*. Hence when Newton characterizes forces as causes of changes in states of motion, he typically does not elaborate. For instance, in the Scholium, he writes: "The causes which distinguish true motions from relative motions are the forces impressed upon bodies to generate motion" (*Principia*, 412). So forces are causes in an ordinary sense.

There are at least two distinct ways of thinking about forces, according to Newton. Near the opening of the *Principia* he contrasts what he calls the "mathematical" and the "physical" treatment of force.[21] In the definitions, after defining various sorts of motion and of force, and in particular after defining what he takes to be the various measures of centripetal force, Newton writes of the concept of force as he employs it in general: "This concept is purely mathematical, for I am not now considering the physical causes and seats of these forces" (*Principia*, 407). Similarly, in the Scholium

---

[20] To be more precise, I refer throughout this chapter to what Newton calls impressed forces. He also famously discusses what he calls the *vis inertiae*, or sometimes the *vis insita*, of material bodies, but this is not an impressed force in his sense and therefore must be treated separately. See the lucid discussion of this point in Cohen, "Newton's Copy of Leibniz's *Théodicée*," 410–11 n. 3.

[21] See the extensive discussion in Cohen, *The Newtonian Revolution*, 52–156.

following section 11 of book I, he describes his use of the term "impulse" by noting that he considers "not the species of forces and their physical qualities but their quantities and mathematical proportions, as I have explained in the definitions" (*Principia*, 588). So whereas a physical treatment of force describes, among other things, its "species" and "physical qualities," a mathematical treatment eschews such a description, providing instead a characterization of its "quantities." But what does this mean? And does Newton ever discuss the "species" of forces and their "physical qualities"? These questions will guide the discussion below.

But there are more general issues to be addressed. As a first step toward clarification, I suggest a caveat: we cannot interpret the distinction between mathematical and physical treatments of force as a straightforward one. I propose instead that we think of the distinction as a technical one; by calling it technical, I mean that the distinction cannot be understood antecedently to its articulation in the text, despite the fact that it appears to be a perfectly familiar distinction. Yet it is not technical in either of the two ways mentioned above, for: (1) it does not involve any explicit definitions; and (2) there is no distinction between an ordinary and a mathematical conception of force, for Newton apparently thinks there is no ordinary conception of force, unlike space or time. Instead, one must read through both books I and III of the *Principia* in order to clarify what the distinction means. The danger is that it may seem straightforward to distinguish the mathematical and the physical. For instance, we might say that triangles or numbers are mathematical entities, just as planets or tables are physical ones. Yet as we will see, when Newton analyzes forces in a physical way in the *Principia*, he is not dealing with physical objects or properties in the usual sense. Forces are not obviously properties of material objects, nor are they obviously objects in their own right. This represents a hallmark of Newton's approach: he has a way of analyzing forces that is neutral on what we might take to be obvious questions about the ontology of forces.

But if forces are not obviously properties or objects themselves, what are they? Answering this question will, it is hoped, involve a second step toward clarification. Newton's point that the mathematical treatment of force focuses on a force's "quantities," rather than its physical "species" or physical "cause," is more illuminating than it might appear. In a crucial sense, the *Principia* concerns quantities throughout. As Newton makes perfectly clear, he takes all of the following to be quantities: force, mass, space, time, and motion (including its derivatives, such as quantity of motion, or momentum). We can measure mass, acceleration, and force by virtue of their relationship with one another. So regardless of any other questions regarding the ontology of force – some of which I discuss below – Newton's mathematical treatment of force indicates how to measure a force by measuring mass and acceleration. This is essential to Newton's approach. We can think of forces as physical quantities, precisely as the "quantity of matter" – i.e. mass – is a physical quantity. They are physical quantities because they are part of the physical world; and they are part of the physical world because they can be measured by measuring other, obviously physical, quantities. So my suggestion will be that the mathematical treatment of force measures physical quantities. Hence it is not mathematical in the sense that it deals solely with mathematical entities. And as we will see, the physical treatment of force is not defined as dealing with physical entities, at least in the usual sense; instead, each of them deals with physical quantities. The mathematical analysis indicates how to measure such quantities; as we will see below, the physical analysis uncovers the "species" and "cause" of the forces in question. This is the beginning of Newton's answer to the question: what does it mean to claim that forces in general, or that gravity in particular, exist? It means that a physical quantity that we can measure exists.

But what, then, is the "species" of a force? Before we can answer this question, we need a third step toward clarification, one involving the structure of the *Principia* itself. We can interpret book I of the *Principia* as providing a merely mathematical treatment of force, and book III as providing aspects of a physical treatment of force, especially of the force of gravity. Newton himself hints at this point in the preface that opens book III:

In the preceding books I have presented principles of philosophy that are not, however, philosophical but strictly mathematical – that is, those on which the study of philosophy can be based. These principles are the laws and conditions of motions and of forces, which especially relate to philosophy ... It still remains for us to exhibit the system of the world from these same principles. (*Principia*, 793)

On the basis of the mathematical principles outlined in book I, including the laws of motion, Newton then analyzes the "system of the world" – our solar system – in book III, arguing that a particular force, namely universal gravity, is especially salient for understanding the motions of the bodies in that system. Whereas book I considers various possible motions of bodies without regard to the actual motions of bodies in the solar system, or anywhere in nature, book III then attempts to determine, first, what the actual motions of bodies in the solar system are, and then second, to determine what force, or forces, cause those motions.[22]

This procedure is clearly articulated in the author's preface to the first edition of the *Principia*. His text, he writes, sets forth

mathematical principles of natural philosophy. For the basic problem of philosophy seems to be to discover the forces of nature from the phenomena of motions and then to demonstrate the other phenomena from these forces. It is to these ends that the general propositions in book 1 and 2 are directed, while in book 3 our explanation of the system of the world illustrates these propositions. For in book 3, by means of propositions demonstrated mathematically in books 1 and 2, we derive from celestial phenomena the gravitational forces by which bodies tend toward the sun and toward the individual planets. Then the motions of the planets, the comets, the moon, and the sea are deduced from these forces by propositions that are also mathematical. (*Principia*, 382)

The *Principia* is mathematical in part because the physical treatment of force in book III is based on the principles outlined in book I, including geometric propositions and the laws of motion themselves; it is a work in natural philosophy because those mathematical principles are employed in an analysis of the actual motions of natural bodies, especially those in our solar system. In their correspondence in 1686, Newton and Halley (who saw the first edition through the press) emphasize that book III's analysis of the motions in the solar system is crucial to calling the whole *Principia* a work in natural philosophy.[23]

We are now in a position to understand what Newton means by the "physical species" of a force. Consider, first of all, Newton's most famous discussion of these matters, in the Scholium after the definitions:

Further, it is in this same sense that I call attractions and impulses accelerative and motive. Moreover, I use interchangeably and indiscriminately words signifying attraction, impulse, or any sort of propensity toward a center, considering these forces not from a physical but only from a mathematical point of view. Therefore,

---

[22] For an illuminating discussion of related issues, see especially Smith, "The Methodology of the *Principia*." In "Newton and the Reality of Force," I do not indicate clearly that book III of the *Principia* provides essential components of a physical treatment of force, especially of the force of gravity. In particular, I unjustly ignore the fact that Newton regards the law of universal gravitation as specifying what he calls the "physical species" of gravity. In the interim, I have been influenced by many discussions with George Smith.

[23] As I mention above, on 20 June 1686, Newton wrote to Halley as follows: "The two first books without the third will not so well bear the title of *Philosophiae naturalis Principia Mathematica* & therefore I had altered it to this *De Motu corporum libri duo*: but upon second thoughts I retain the former title" (*Correspondence*, vol. II: 437). Two weeks earlier, Halley had written to Newton that he ought to include book III because "the application of this mathematical part, to the system of the world; is what will render it acceptable to all naturalists, as well as mathematicians, and much advance the sale of the book" (*Correspondence*, vol. II: 434).

let the reader beware of thinking that by words of this kind I am anywhere defining a species or mode of action or a physical cause or reason, or that I am attributing forces in a true and physical sense to centers (which are mathematical points) if I happen to say that centers attract or that centers have forces. (*Principia*, 408)

Now in the definitions that precede book I, Newton defines "impressed force" as: "the action exerted on a body to change its state either of resting or of moving uniformly straight forward" (*Principia*, 405). There are at least three kinds of impressed force: percussion, pressure, and (what he famously calls) centripetal force. The latter, of course, is Newton's focus, and he then defines centripetal force as "the force by which bodies are drawn from all sides, are impelled, or in any way tend, toward some point as to a center" (*Principia*, 405). In the next three definitions – six, seven, and eight – Newton provides three measures for centripetal force, the motive, the accelerative, and the absolute measure (*Principia*, 405–8). These provide us with three ways of measuring a centripetal force.

These are essential aspects of a mathematical treatment of force for at least two reasons. First, we can measure a centripetal force by measuring the velocity or the motion that it generates in a given time in some body; for instance, we can measure the effect that a magnet has on some iron filings. But we ignore various other questions: we bracket the question of whether the force is propagated through a medium, whether it occurs only through impacts between macroscopic bodies, or whether it may involve microscopic particle interactions, etc. That is, we are not considering, as Newton says in definition eight, "the physical causes and sites of forces." We can measure a centripetal force without considering whether it is caused by (say) the magnet itself, by some medium between the magnet and the iron, or through some other means that we cannot yet envision. Second, we can measure a centripetal force, or indeed any force, without analyzing the actual motions of the bodies in nature, and therefore without analyzing what kinds of forces might produce those actual motions.

The difference between books I and III, and between the mathematical and the physical treatment of force, can then be seen in the discussion appended to definition five, the definition of centripetal force quoted above:

One force of this kind is gravity, by which bodies tend toward the center of the earth; another is magnetic force, by which iron seeks a lodestone; and yet another is that force, whatever it may be, by which the planets are continually drawn back from rectilinear motions and compelled to revolve in curved lines. (*Principia*, 405)

In book III, which takes account of the actual motions in the solar system – described as the six "phenomena" that open the book (*Principia*,

797–801) – Newton then reaches a startling conclusion. In the Scholium to proposition 6, he harkens back to the discussion appended to definition five, writing: "Hitherto we have called 'centripetal' that force by which celestial bodies are kept in their orbits. It is now established that this force is gravity, and therefore we shall call it gravity from now on" (*Principia*, 806). We are no longer dealing purely mathematically with a centripetal force, as we were in book I; we are discussing a particular "species" of centipetal force. And this is a crucial component of the physical treatment of gravity.

But what does it mean to say that Newton discovers the "species" of centripetal force that causes the motions in our solar system, and how is that part of the physical treatment of force? The investigation of gravity centers on the discovery of a force law, which of course is the highlight of book III of the *Principia*, where Newton "derives" the law of universal gravitation. The law that governs gravity indicates its "species," which indicates, in turn, how gravity differs from other members of the genus, centripetal force, such as magnetism. Newton makes this point in corollary five to proposition 6 of book III:

The force of gravity is of a different kind from the magnetic force. For magnetic attraction is not proportional to the [quantity of] matter attracted. Some bodies are attracted [by a magnet] more [than in proportion to their quantity of matter], and others less, while most bodies are not attracted [by a magnet at all]. And the magnetic force in one and the same body can be intended and remitted [i.e. increased and diminished] and is sometimes greater in proportion to the quantity of matter than the force of gravity; and this force, in receding from the magnet, decreases not as the square but almost as the cube of the distance, as far as I have been able to tell from certain rough observations. (*Principia*, 810)

Once we are able to determine that the magnetic force does not follow the law of universal gravitation, we are able to determine that it is a distinct "species" from gravity. Since a force just is a physical quantity, its "physical species" is determined solely by the law that enables us to measure it. A bit more precisely, the law of universal gravitation says

F_{grav} is proportional to $m_A m_B / r^2$

where m_A and m_B are the masses of two bodies and r is the distance between them.[24] This law allows us to measure the force of gravity between two

---

[24] This is closer to Newton's formulation than the modern version of the law, which says:

$$F_{grav} = G\, M_A M_B / r^2$$

because Newton used proportionalities and did not think in terms of the gravitational constant. Thanks to George Smith for discussion of this issue.

massive bodies by measuring other physical quantities, such as the distance between them. This indicates the "physical species" of gravity.

We are now in a position to see why a work in natural philosophy ought to include what Newton calls a physical treatment of force, and not only a mathematical treatment. As we saw in the case of centripetal force above, the mathematical treatment in book I enables us to measure any centripetal force through a number of means, thereby enabling us to think of that force as a physical quantity, one that differs from other impressed forces, such as percussion, that do not involve a motion of a body toward some center. This treatment is mathematical because it involves the measurement of this quantity; but it is merely mathematical because it is agnostic on the question of whether there are any centripetal forces in nature. It ignores the actual motions of bodies in the solar system, and it ignores the question of whether there may be distinct "species" of centripetal force in nature, for instance, gravity and magnetism. But as we have seen, natural philosophy crucially involves, for both Halley and Newton already in 1686, an analysis of the actual motions of bodies in nature – thus it is only with the physical treatment of book III in the *Principia* that we have something beyond a merely mathematical treatise on motion in general. Halley and Newton concurred that without book III, the text would be better titled *De Motu corporum*. Yet the work in book III remains mathematical because it involves the same techniques for measuring centripetal forces; it then adds an analysis of the motions in nature, allowing us to trace them back to a particular "species" of centripetal force.

Yet Newton's physical treatment of force in book III is avowedly incomplete. It indicates the "physical species" of the force causing various motions throughout the solar system, but famously fails to determine gravity's "physical cause or reason." That is, Newton's theory does not indicate whether gravity depends in some way on the ether or some other medium that fills the solar system, or perhaps all of space. After discussing the actual motions of the bodies in the solar system, Newton concludes that it is universal gravity – as measured by his law – that causes those motions, and not any other "species" of force, such as magnetism or electricity or the like. This includes the stunning claim that the free fall of bodies on earth, the motions of the moon, and the planetary orbits, are all caused by one and the same force, governed by one and the same law. Yet this conclusion does not settle the crucial question of the "physical cause or reason" of gravity. Does the pressure of an ether cause these motions? Or are they due to a shower of imperceptible particles? Or something else not yet imagined? It is these questions, in turn, that seem the most relevant for determining what really causes the actual motions of the bodies in our world. If Newton is ignorant of gravity's "physical cause," what can it mean to say, as he does, that gravity really exists and that it causes the motions in question? These questions remain.

Before attempting to clarify my interpretation in what follows, I want to indicate what is at stake in the proper interpretation of Newton's "mathematical point of view" by considering its place in the work of his most prominent critic and one of his most prominent defenders, historically speaking. Both Leibniz and Clarke invoke Newton's distinction between the mathematical and the physical treatment of force in the course of pressing their cases against one another, and the question of the place of causation within the mathematical treatment of force is central for each of them.

FORCES IN LEIBNIZ AND CLARKE

Among Clarke's myriad defenses of Newtonian natural philosophy in the face of Leibnizian criticisms, we find the argument that the *Principia*'s treatment of force can be construed as involving a kind of causal agnosticism, one that allows Newton to avoid the charge that he relies on action at a distance. Significantly, in interpreting Newton's mathematical treatment of force, Leibniz and Clarke each appear to be satisfied with employing an ordinary conception of causation in order to present their differing characterizations of Newton's causal commitments. That is to say, each deploys an ordinary notion when determining whether Newton is committed to thinking of gravity as a cause of certain phenomena. So their debate is not merely semantic.

Having read Leibniz's constant criticisms of Newton's theory of gravity, in his last letter to Leibniz Clarke insists:

It is very unreasonable to call attraction a miracle, and an unphilosophical term; after it has been so often distinctly declared, that by that term we do not mean to express the cause of bodies tending toward each other, but barely the effect, or the phenomenon it self, and the laws or proportions of that tendency discovered by experience; whatever be or be not the cause of it.[25]

And then in the same letter he writes:

The phenomenon itself, the attraction, gravitation, or tendency of bodies towards each other (or whatever other name you please to call it by) and the laws, or

---

[25] See sections 110–16 of Clarke's fifth letter in Leibniz, *Die philosophischen Schriften*, vol. VII: 437–8.

proportions, of that tendency, are now sufficiently known by observations and experiments. If this or any other learned author can by the laws of mechanism explain these phenomena, he will not only not be contradicted, but will moreover have the abundant thanks of the learned world.[26]

So the *Principia's* treatment of gravity precisely describes certain tendencies to motion, but remains neutral as to the cause of those tendencies. He construes gravity as a "matter of fact" concerning various motions, rather than as a cause.

Clarke's attempted defense of Newton has an intriguing and influential echo in the contemporary literature, especially in two recent papers by Ernan McMullin in which he presents an illuminating interpretation of Newton's treatment of force.[27] McMullin notes that there is a tension between Newton's use of terms like "attraction" and his caveats about those terms, claiming that "attraction" can be given two distinct meanings in the *Principia*.[28] The meanings are as follows:

(1) in contending, for instance, that the sun "attracts" the earth, the sun is conceived of as an "agent"; i.e. the sun is construed as the cause of the earth's motion;

(2) the contention that the sun "attracts" the earth is construed to mean only that the earth has a "disposition" to move in various ways.

It is natural, McMullin thinks, to interpret Newton's eschewing of the "physical" treatment of gravity as a rejection of (1) in favor of (2), which can be construed as a "merely mathematical" treatment. McMullin is, of course, quite correct, as we have seen, that Newton employs (e.g.) the terms "attract" and "attraction" at various stages of his argument; McMullin's view is that to provide a merely mathematical understanding of the use of such terms is to construe them as referring to dispositions to motion.

Interpreting the dispositional construal of the mathematical treatment of force requires care, because one does not want it to represent a way of settling the question of gravity's "physical cause or reason." If we construe the claim that the sun "attracts" the earth to mean that the earth has a disposition to follow a particular elliptical orbit with the sun lying at one focus of the ellipse, and we think of this as a way of avoiding the ascription of action at a distance to any of the bodies in question, we can avoid any characterization of gravity's physical cause if we do not take this account to

---

[26] See sections 118–23 of Clarke's fifth letter in Leibniz, *Die philosophischen Schriften*, vol. VII: 439–40; cf. sections 110–16 of the same letter, 436–8.

[27] See McMullin's essays, "The Impact of Newton's *Principia* on the Philosophy of Science," and "The Origins of the Field Concept in Physics."

[28] See, for instance, McMullin, "The Origins of the Field Concept in Physics," 23.

rule out various possibilities, such as the possibility that the earth is continually pushed by an ether along its orbit. Perhaps the earth would be pushed in virtue of its having mass, so its disposition to motion would be a relational property dependent on its mass and on the characteristics of the ether.[29]

The point is that by Newton's lights, the use of the terms "attraction" and "attract" is intended to be compatible with any physical account of gravity in the sense that it is not intended to rule out any physical medium that "pushes" or "pulls" bodies, as long as that medium's properties and modes of action are consistent with the law of universal gravitation (I discuss the issue of consistency below). Of course, that is perfectly compatible with the claim that our ignorance concerning the ether's operation – assuming for a moment that we have empirical evidence indicating the existence of an ether, but no understanding of how it operates on bodies – might be expressed by contending that it is the ether that has a disposition to push, or pull, the planetary bodies in various directions. That is certainly a way of expressing agnosticism about the physical operation of the ether, but it is not, it seems to me, a way of expressing the type of agnosticism signaled in Newton's contention that he leaves open what gravity's physical cause might be.

McMullin's own construal of the dispositional account of the mathematical treatment of force may press us in a slightly different direction than the one I have just outlined. After noting that Newton was committed to rejecting action at a distance *per se*, he discusses an implication of this rejection:

The significance of this point for my theme is that it means that the "mathematical" or "dispositional" interpretation of the mechanics of the *Principia* was, in Newton's own eyes, incomplete; in the long run it would have to be supplemented by a properly "physical" account of just how the active disposition of one body could affect the state of motion of another causally at a distance from it. The only reading of the mathematical-dispositional account of gravity that would transform it directly into a physical account, i.e. that would allow it to pose as complete, would necessarily limit it to a postulation of unmediated action at a distance. And this Newton did not want. (McMullin, "The Origins of the Field Concept in Physics," 24)

---

[29] For discussions of Newton's understanding of the ether, see Cantor and Hodge's introduction to *Conceptions of Ether*, 22–30, Aiton, *The Vortex Theory of Planetary Motions*, 108ff., and Maxwell, *The Scientific Papers*, 764.

As will become clear below, I concur with McMullin's (perhaps surprising) contention that the mathematical account may have to be supplemented with, but not replaced by, a physical account of gravity. But his way of characterizing the dispositional account in the passage above indicates that it already rules out any medium between bodies to account for gravitational phenomena like the planetary orbits, for the claim that the earth and the sun are causally distant expresses the fact that we know there to be no causally efficacious medium between them. And according to the construal above, to assert that the earth is "disposed" to orbit the sun is *ipso facto* to assert that the two are causally distant. But from Newton's point of view, that is precisely the sort of proposition that we do not know.

Of course, this concerns only McMullin's gloss on his own dispositional construal of Newton's treatment of force. If we do not endorse that gloss, we are left with something akin to Clarke's view: the use of terms such as "attraction" describes only dispositions or tendencies to motion and is in fact causally neutral; it is neutral in the sense that it is silent on what leads to the various motions of the heavenly bodies. It is precisely this interpretation of Newton's commitments that Leibniz rejects.

One of Leibniz's principal tactics is to praise the *Principia*'s treatment of force, but then to contend that Newton ultimately strays from its safe harbor into philosophically problematic territory. To his credit, in the *Tentamen* of 1689 – which represents one of Leibniz's most extensive responses to the *Principia*, and which outlines a vortex theory of planetary motion – Leibniz acknowledges the significance of Newton's mathematical achievement in the *Principia*.[30] He even notes that the planets are "attracted" by the sun as $1/r^2$, something that was apparently "also known" to the "renowned Newton."[31] Several years later, when writing to Newton directly in 1693, Leibniz makes a related acknowledgment, while expressing his commitment to his own vortex theory of gravity:

You have made the astonishing discovery that Kepler's Ellipses result simply from the conception of attraction or gravitation and passage in a planet. And yet I would incline to believe that all these are caused or regulated by the motion of a fluid medium, on the analogy of gravity and magnetism as we know it here. Yet this solution would not at all detract from the value and truth of your discovery. (Leibniz, *Die mathematische Schriften*, vol. I: 169)

---

[30] Newton wrote some short, and somewhat tendentious, notes on Leibniz's *Tentamen* – they are available in *Correspondence*, vol. VI: 116–22.

[31] Leibniz, *Die mathematische Schriften*, vol. VI: 157, and Bertoloni Meli, *Equivalence and Priority: Newton vs. Leibniz*, 138. Since the latter includes the standard edition of Leibniz's *Tentamen* in English, I follow its renderings of the Latin.

The obvious question here is this: can Leibniz genuinely acknowledge that the planetary orbits are due to "attraction"? Could this be anything more than a disingenuous remark intended to keep the peace? I will return to that issue shortly.

Unlike Clarke, Leibniz did not allow Newton's account of gravity in the *Principia* to be interpreted only as a causally neutral theory of force, that is, as a mathematically precise description of certain motions, or dispositions to motion, that is agnostic on their cause. He consistently understood Newton as committed to the view that the world actually contains an "attractive" force, i.e. one that is causally efficacious without contiguity because it acts independently of any medium, whether vortex-like or otherwise. Leibniz often bemoaned this deviation from the contemporaneous philosophical consensus.[32] As he insists in his last letter to Clarke:

> For it is a strange fiction [*étrange fiction*] to make all matter gravitate, and that toward all other matter, as if all bodies equally attract all other bodies according to their masses and distances, and this by an attraction properly so called [*une attraction proprement dite*], which is not derived from an occult impulse of bodies, whereas the gravity of sensible bodies toward the center of the earth ought to be produced by the motion of some fluid. And it must be the same with other gravities [*d'autres pesanteurs*], such as is that of the planets toward the sun, or toward each other. A body is never moved naturally except by another body that touches it and pushes it; and after that it continues until it is prevented by another body that touches it. Any other operation on bodies is either miraculous or imaginary. (*Die philosophischen Schriften*, vol. VII: 397–8)

So Leibniz reads Newton as presenting a twofold description of gravity. On the one hand, when writing strictly from a mathematical point of view, Newton construes "attraction" agnostically to express the fact that it is as if the sun attracts the earth, or the earth the moon, etc. This construal does not express what actually causes the motions of these bodies toward one another. This view, according to Leibniz, is perfectly defensible because it does not take attraction to name any existing force. In that sense Leibniz's praise for Newton in his 1693 letter was genuine.

On the other hand, Newton also insists on articulating a causal claim concerning the phenomena associated with gravity – he does so, for instance, in his response to Leibniz's 1693 letter, as we will see below – and in so doing, he transcends the limits of a merely mathematical

---

[32] Later in life, during the height of the calculus priority dispute, Newton had become particularly sensitive to Leibniz's contention that the *Principia* postulated action at a distance between material bodies. For instance, when reading Leibniz's *Theodicy* of 1710, Newton marked the passage in which Leibniz made this criticism – see Cohen, "Newton's copy of Leibniz's *Théodicée*," 411.

treatment of gravity. He wrongly attributes the motions associated with gravity to attraction properly so-called; that is, he takes attraction to name an existing force and thereby invokes action at a distance.[33] In his 1693 letter to Newton, Leibniz insists that a mathematical construal of "attraction" should not press us into invoking a genuine attraction. Rather, for the relevant causal chain, we must look to the motion of a vortex, thereby indicating that the apparent attraction is reducible without remainder to contact action of a familiar and acceptable variety. But, of course, Newton consistently dismisses any vortex theory of gravity *à la* Leibniz.[34]

The question we now face, therefore, is whether Clarke's attribution of causal agnosticism to Newton is defensible in the light of Leibniz's contention that Newton spoils his mathematical account of attraction by rejecting vortices and consistently invoking "attraction properly so-called." On Leibniz's reading, Newton takes a stand on the "physical cause or reason" of gravity, claiming that it is the masses of, and the distances between, material bodies independently of any medium between them. The claim that Newton is genuinely committed to the view that gravity itself is the cause of the planetary orbits is buttressed by the fact that Newton viewed his theory as a competitor to a type of causal theory that was prevalent at the time the *Principia* first appeared, the theory that gravity is propagated through some kind of vortex.[35] Leibniz himself, of course, consistently defended such a theory, as Newton well knew, so we can gauge the depth of Newton's causal commitment by considering his response to the type of theory Leibniz defends in the *Tentamen*. The *Tentamen*, in turn, is an essential component in Leibniz's response to the *Principia*.

As we saw in ch. 2, in the *Tentamen* Leibniz emphasizes that prior to any empirical research, we already know both the nature of motion and the nature of bodies. Moving bodies tend to recede along the tangent to any curve, and the state of motion of any given body can be altered only by something "contiguous" to that body. Hence there can be no action between distant bodies. If we begin from the assumption that the planets follow curvilinear paths around the sun, it follows from the nature of

---

[33] As I discuss in ch. 6 below, Cotes raised this issue with Newton as he was editing the second edition of the *Principia* in 1713.

[34] He does so in part because of the "retrograde" motion of some comets – such as Halley's comet – for they enter the solar system at various angles and traverse the planetary orbits against what would presumably be the rotational motion of the planetary vortices. Hence the vortices ought to impede the motion of such comets. Thanks to George Smith for discussion of this point.

[35] For discussion, see Bertoloni Meli, *Equivalence and Priority*, 38–50.

motion that something must intervene to prevent them from following the tangents to those paths, and it follows from the nature of bodies that whatever alters their motion in this respect must be "contiguous" to them.

Leibniz's argument then proceeds as follows. We introduce, *ex hypothesi*, the claim that a fluid surrounds, and is contiguous to, the various planetary bodies, and we then argue that this fluid must be moving.[36] Venturing beyond an agnostic treatment by proposing a causal analysis of the planetary orbits must involve an attribution of contact action between the planets themselves and some physically characterized entity or medium, such as a vortex, that is contiguous to the planets. This is a bedrock assumption for Leibniz.

Newton was perfectly familiar with this Leibnizian account of gravity; how did he respond to it? He obviously accepted aspects of it, such as the principle of inertia, although his understanding of inertia differs from Leibniz's: Newton tended to think that it expressed, not the nature of motion, but the nature of matter, though little seems to hinge on that difference here.[37] And Leibniz speaks of motion rather than the state of motion, which may indicate his failure to appreciate an implication of the principle of inertia. Newton would also object to Leibniz's use of an *ex hypothesi* premise, which introduces the claim that the bodies in question – the planets – sit in a fluid. Newton would complain that there is no independent empirical evidence for the existence of such a fluid. The disagreement, therefore, would partially involve Leibniz's willingness to proceed in this avowedly "hypothetical" manner.[38]

But the disagreement would not end there, and this is particularly salient for our purposes here. Just four years after the *Tentamen* was written in

---

[36] Leibniz writes: "To tackle the matter itself, then, it can first of all be demonstrated that according to the laws of nature *all bodies which describe a curved line in a fluid are driven by the motion of the fluid*. For all bodies describing a curve endeavor to recede from it along the tangent (from the nature of motion), and it is therefore necessary that something should constrain them. There is, however, nothing contiguous except for the fluid (by hypothesis), and no conatus is constrained except by something contiguous in motion (from the nature of the body), therefore it is necessary that the fluid itself be in motion." Leibniz, *Die mathematische Schriften*, vol. VI: 149, and Bertoloni Meli, *Equivalence and Priority*, 128–9.

[37] On Newton's acceptance of the first two premises in Leibniz's argument, see De Gandt, *Force and Geometry in Newton's "Principia,"* 269–70. Concerning his view that inertia expresses the nature of bodies rather than the nature of motion, see Newton, *Principia*, 796, and cf. also the illuminating discussion in Stein, "Newton's Metaphysics," 289.

[38] The literature on Newton's attitude toward hypotheses is enormous; see the classic account in Cohen, "Hypotheses in Newton's philosophy," 163–84, and the brief discussion in my introduction to *Philosophical Writings*, xxiv–xvi. See also ch. 2 above. On the role of hypotheses in Cartesian natural philosophy, which is obviously relevant for understanding Newton's attitude toward them, see Clarke, *Occult Powers and Hypotheses*, 141–90.

1689, Newton responded to Leibniz's letter of 1693 with a rather startling claim, one which Leibniz clearly had in mind when contending that Clarke failed to acknowledge Newton's commitment to causal claims regarding the phenomena associated with gravity. In his response to Leibniz, Newton contends that the *Principia* indicates the following:

> For since celestial motions are more regular than if they arose from vortices and observe other laws, so much so that vortices lead not to the regulation but to the disturbance of the motions of planets and comets; and since all phenomena of the heavens and the sea follow precisely, so far as I am aware, from nothing but gravity acting in accordance with the laws described by me [*cumque omnia caelorum et maris phaenomena ex gravitate sola secundum leges a me descriptas*]; and since nature is simple, I have myself concluded that all other causes are to be rejected and that the heavens are to be stripped as far as may be of all matter, lest the motions of planets and comets be impeded or rendered irregular. (*Correspondence*, vol. III: 286)

The claim is that vortices cannot be the cause of the planetary orbits – for complex but reasonably well-known reasons that I will not delve into here[39] – and that gravity alone should be singled out as their cause. Newton's causal view, therefore, is an integral component of an ongoing program of gravitational research, a program that he uses to rebut the vortex program favored by Leibniz.

This perspective appears to represent Newton's considered view: it is apparent, for instance, in his much more vociferous debate with Leibniz in the early 1710s. In May of 1712, Leibniz wrote to Nicholas Hartsoeker to present several familiar criticisms of the Newtonians; his letter was later published in English in the *Memoirs of Literature*. After learning of the letter from Roger Cotes, the editor of the second edition of the *Principia* and a subscriber to the *Memoirs*, Newton wrote an only posthumously published rebuttal in which he rejected the Leibnizian criticism that his theory rendered gravitation a "perpetual miracle" because it failed to specify what Leibniz would countenance as a relevant physical mechanism. Newton first paraphrases a Leibnizian criticism, one repeated in the Clarke correspondence:

---

[39] See the discussions in Aiton, *The Vortex Theory of Planetary Motions*, 110ff., and Bertoloni Meli, *Equivalence and Priority*, 191–218. See also query 28 in *Opticks*, and Cotes's preface to the 1713 edition of the *Principia*, in which he outlines some of Newton's reasons for rejecting vortices – *Principia*, 385–99.

But he [Leibniz] goes on and tells us that God *could not create planets that* should move round of themselves without any cause that should prevent their removing through the tangent. For a miracle at least must keep *the planet in.*

Newton's rebuttal is illuminating:

But certainly God could create planets that should move round of themselves without any other cause than gravity that should prevent their removing through the tangent. For gravity without a miracle may keep the planets in.[40]

So Newton repeats his assertion from his 1693 letter to Leibniz that gravity itself causes the planets to follow their orbital paths instead of their inertial trajectories along the tangents to those orbits.

Newton therefore consistently asserts in various pieces of correspondence between 1693 and 1712 that gravity – as he understands it – is to be interpreted as a genuine cause, as an existing force. Even more importantly this view is reflected in the *Principia* itself. In the definitions that open the *Principia*, Newton characterizes the two forces that he will later identify with one another via rule two (see below) in a particularly salient way. In definition five he defines "centripetal force" as the force by which bodies tend toward a point as a center (*Principia*, 405). Two examples of this force are terrestrial gravity, and whatever force compels the planets to retain their solar orbits. In definition four we learn that centripetal forces are sources of impressed force, and the latter is an action that alters the state of motion of any body. As he writes later, in the Scholium: "The causes which distinguish true motions from relative motions are the forces impressed upon bodies to generate motion" (*Principia*, 412). So Newton conceives of centripetal forces as causes in an ordinary sense: they alter the states of motion of material bodies. Thus when he identifies these two examples of centripetal force together in book III Newton clearly conceives of the force of gravity itself as altering the states of motion of material bodies.

Moreover, Newton explicitly employs the so-called rules of reasoning in later editions of the *Principia* to derive, in book III, the universality of gravity. For instance, consider rule two, which says: "Therefore, the causes assigned to natural effects of the same kind must be, so far as possible, the same" (*Principia*, 795). This rule is employed to identify the cause of the weight of objects near the surface of the earth – which Newton calls "gravity" throughout the *Principia*, before he has derived universal

---

[40] For both quotes from Newton's letter, see *Correspondence*, vol. V: 300; the letter is reprinted in *Philosophical Writings*, 114–17, and discussed briefly in my introduction to that volume, xxviii–xxxi. This exchange occurred while Cotes and Newton were preparing the changes to be incorporated in the second edition of the *Principia*.

gravity – and the force that maintains the planets in their solar orbits, and the moon in its terrestrial orbit. He contends that these are the same effects, and thus, by rule two, they ought to be assigned the same cause.[41] That is to say, Newton's use of his own methodological principles commits him explicitly to thinking of gravity as a genuine cause.

We are therefore left with the following question: assuming that Clarke's defense of Newton is not viable, and that Leibniz is correct in claiming that Newton is committed to the view that the force of gravity exists – each of which I discuss below – how can Newton hope to avoid an entanglement with action at a distance? And how can Newton insist that gravity itself causes the motions of the bodies in the solar system when he remains ignorant of its physical cause? After all, perhaps the ether, or some other medium, does all the causal work.

NEWTON'S DILEMMA RESOLVED

Given the specter of action at a distance, and given Leibniz's powerful criticism of Newton's text, what possible meaning can Newton's claim that gravity "really exists" retain? It seems to me that Newton's claim has at least the following three meanings.[42] Discussing them will once again raise the question of how we should understand Newton's ignorance of gravity's "physical cause or reason."

First, it means that a wide range of previously disparate phenomena – including the free fall of bodies on earth, the tides, and the planetary and satellite orbits – have the same cause. It is precisely to highlight this startling result, of course, that we call this cause gravity; indeed, before reaching this conclusion in book III of the *Principia*, Newton employs such phrases as "whatever force maintains the planetary bodies in their orbits." The culmination of Newton's argument is the identification of this force as "gravity." So although the use of the term "gravity" is neutral with respect to the physical characterization of the cause of these phenomena in Newton's technical sense, it is not neutral on the question of whether the free fall of some rock toward the surface of the earth, and the orbital path of some extremely distant comet, have the same cause. Newton says they do.

---

[41] See *Principia*, 391 and 806, and De Gandt, *Force and Geometry in Newton's "Principia,"* 266.

[42] The discussion here expands the necessarily limited treatment in ch. 2. See also the helpful discussion in Broughton, "Hume's ideas about necessary connection," 233–4; her interpretation is compatible with my own in certain respects.

The second meaning is this: the various phenomena caused by gravity are such that mass and distance are the only salient variables in the causal chain that involves them. We express this precisely through the law of universal gravitation, asserting that gravity is as the masses of the objects in question and is inversely proportional to the square of the distance between them. This raises several crucial issues, one of which leads us to the third meaning of Newton's contention.

Third, given that mass is one of the salient variables in the causal chain involving the previously disparate phenomena taken by Newton to be caused by gravity, we already know, as Newton himself makes perfectly clear, that gravity is not a mechanical cause. At least we attain such knowledge if we adopt Newton's own understanding of "mechanical" in the *Principia*. As we have seen above, Newton writes in the General Scholium:

Indeed, this force arises from some cause that penetrates as far as the centers of the sun and the planets without any diminution of its power to act, and that acts not in proportion to the quantity of the *surfaces* of the particles on which it acts (as mechanical causes are wont to do) but in proportion to the quantity of *solid* matter, and whose action is extended everywhere to immense distances, always decreasing as the squares of the distances. (*Principia*, 943)

Newton's important contention that gravity is not a mechanical cause in a significant sense – that it does not act "as mechanical causes are wont to do" – can easily be interpreted as undermining his claim to have avoided action at a distance, as Leibniz would surely insist.[43] So problems remain.

To understand this third aspect of Newton's contention we must rely on an interpretive maneuver employed several times above: Newton's use of the word mechanical here is a technical one.[44] He effectively accused some defenders of the mechanical philosophy, including Leibniz, of conflating local with surface action (or impact). According to the overarching view that Newton would attribute to Leibniz, a cause must involve some mechanism – it must be "mechanical" – in the following two senses: (1) the cause cannot

---

[43] In a draft of possible alterations to the second edition of corollaries four to five of proposition 6 to book III, Newton writes that any hypothesis by which one purports to explain gravity "mechanically" must somehow account for the fact that it is proportional to mass – see Hall and Hall, *Unpublished Scientific Papers of Isaac Newton*, 315. The Halls date this draft to the early 1690s, a time when Newton first prepared possible alterations for the second edition of the *Principia*; as it turns out, that edition did not appear until 1713. In reflecting on Newton's theory of gravity, Samuel Clarke makes a somewhat stronger claim, contending that gravity is not mechanical (in the sense outlined above) and that matter could only act on surfaces – see his Boyle Lectures of 1704, *A Demonstration of the being and attributes of God*, 58.

[44] In reviewing this aspect of Newton's conception, Maclaurin contends that gravity "seems to surpass mere mechanism" (*An Account*, book IV: 387).

alter the state of motion of any material body at a spatial distance from it; and (2) it cannot alter the state of motion of any material body without impacting on one or more surfaces of that body (recall Leibniz's argument in his *Tentamen*). One might think that the only way to avoid violating condition (1) is to cite causes that meet condition (2). From Newton's point of view, however, this is the conflation of which Leibniz is guilty; (1) and (2) should be seen as distinct. This illuminates Newton's technical sense of mechanical: "mechanical causes" operate only on surfaces. We must reserve room for local action that involves the penetration of a material body by some other body, such as a material particle, or perhaps by another phenomenon, such as a ray of light (leaving aside whether light consists of particles or of waves). This local action would not be mechanical.

The import of this point is not difficult to find. Newton's imagined physical theory of gravity based on the ether in query 21 of the *Opticks* involves an acceptance of (1) but a rejection of (2), for the ether would act on bodies by "penetrating" them.[45] Newton goes so far as to contend, in query 28, that the mechanical philosophers unjustly introduce "hypotheses" into physics precisely by presupposing that physical accounts must meet condition (2); for him, it is an empirical question whether any cause meets that condition (*Opticks*, 368–9). So although all action between material bodies must be local – on pain of there being an "inconceivable" distant action, an issue taken up again in ch. 6 – we cannot presuppose that two macroscopic bodies must interact solely through impact. It may turn out that the constituent microscopic particles of any two bodies do interact by impacting upon one another's surfaces, but we cannot establish this in advance, and it does not follow from this fact that the macroscopic bodies interact with one another via impact. Moreover, the medium may turn out to be continuous rather than particulate, interacting with all the parts of a body, rather than its surface. From Newton's point of view, the development of empirical science has indicated the inadequacy of the position that all causation must be mechanical in his technical sense. This represents an unjust insertion of a metaphysical requirement into physical theory.[46]

To unite these three elements of Newton's view, gravity exists in the following sense. "Gravity" refers to a physical quantity that non-mechanically

---

[45] See *Opticks*, 350–2; cf. the discussion in query 28 at *ibid.*, 368–9.

[46] Newton's overarching view is that some, but presumably not all, causation is non-mechanical. The law of gravity, of course, covers only one force, and even the laws of motion, which can be understood as the most general principles of the *Principia*, cover a restricted range of phenomena, especially objects with mass. Newton knew perfectly well that other phenomena – such as light, electricity, and magnetism – might require a separate analysis.

causes various motions of bodies near the surface of the earth, of our oceans, and of the heavenly bodies, in such a way that distance and mass are the salient variables in their changes in states of motion.[47]

This interpretation should help to illuminate how Newton can justifiably assert that gravity causes various motions while nonetheless avoiding the invocation of action at a distance. The contention that gravity causes the planetary orbits does not amount to, or entail, the contention that there is no causally efficacious medium between planetary bodies that serves as the basis of their gravitational interactions. On the contrary, since Newton's theory is neutral with respect to the physical cause of gravity, it is perfectly compatible with the discovery that some type of medium does in fact exist.[48] Hence, it does not amount to, nor does it entail, the claim that bodies act on one another at a distance through a vacuum; it is in fact entirely compatible with a state of affairs in which all action is local. This is as it should be, for Newton considered any non-local action to be simply "inconceivable."

This interpretation may also help to illuminate Newton's understanding of the treatment of force in the *Principia*, for it allows us to recognize a surprising fact mentioned by McMullin: Newton thinks that any future characterization of gravity's physical cause must somehow cohere with, or account for, the facts established in book III of the *Principia*. In his 1693 correspondence with Leibniz, he explicitly makes this point. After asserting that gravity rather than some combination of vortices causes the planetary orbits – in a passage quoted above – Newton goes on to write:

But if, in the interim, someone explains gravity along with all its laws by the action of some subtle matter, and shows that the motion of planets and comets will not be disturbed by that matter, I shall not object. (*Correspondence*, vol. III: 286)

Newton allows that there might some day be an account of gravity's physical cause involving a "subtle matter," which is to say, not a vortex or some kind of fluid of the sort favored by Leibniz, but some kind of ether. Yet such a future account would not contravene the conclusion that gravity itself is the cause of various phenomena, a claim presented just before the passage above. Why should that be? Because, as we have seen, the claim that

---

[47] For a decidedly different interpretation, one which does not emphasize that gravity is a physical quantity, see McMullin, "The Significance of Newton's *Principia* for Empiricism," 52.

[48] This usage of "cause" is perfectly consistent across texts: for instance, in draft A of the General Scholium, Newton notes that he has not discovered gravity's cause, and he makes it perfectly clear that vortices are not its cause. That is, something like a vortex would be a candidate "cause" for gravity; hence he thinks that a cause is a physical seat (see Hall and Hall, *Unpublished Scientific Papers of Isaac Newton*, 352–3).

gravity causes the planetary orbits should be interpreted to mean that the planetary orbits and the free fall of bodies on earth have the same non-mechanical cause, and that is precisely the surprising datum that a future account of gravity's cause must elucidate.

More specifically, the task of Newton's discussion of the ether in query 21 to the *Opticks* is to indicate how that subtle matter interacts both with bodies near the surface of the earth and with bodies on the other side of the solar system in such a fashion that mass and distance are the salient quantities in their interactions (*Opticks*, 350ff.). The ether could not be mechanical in Newton's sense, but would have to flow through material bodies, interacting somehow with their masses.[49] Similarly, it would have to exhibit differential density, or some other feature that rendered the distance between masses their salient relation. And it would have to exert only a vanishingly small – perhaps undetectable – resistance to the motions of bodies. As Newton knew, no account of gravity's physical cause in this period – including those of Huygens and Leibniz – was remotely capable of accounting for these facts.[50]

Another way to see how the *Principia*'s treatment of gravity and a future account of its cause might cohere with one another is inchoate above, and can be made explicit here: each provides some elements of the complete characterization of the cause of various motions in the world. The *Principia* begins the characterization of the cause by indicating, for instance, how to measure the cause by measuring other physical quantities, and by showing that various motions have the same cause. The account of the physical cause of gravity presupposes and builds on these elements by adding further pieces

---

[49] In the famous second corollary to proposition 6 in book III (*Principia*, 809), Newton rejects a Cartesian material ether on the grounds that it would be a material object that would either be "devoid of gravity," or would not gravitate toward other material objects in proportion to its mass, or "quantity of matter." Following Descartes, Newton envisions here a material ether that would differ from ordinary objects only in its "form," that is, only in the size, shape, and motion of its constituents. That would mean that any piece of matter, if altered in its size, shape, and motion to obtain the form of the ether, could be rendered gravity-less, or could be made to gravitate not in proportion to its mass. And Newton argues here that the weights of all bodies are in fact completely independent of their forms and textures, depending only on their mass. Cf. Boas, "The Establishment of the Mechanical Philosophy," 518.

[50] For broad discussions of Cartesian vortices that include the views of Leibniz and Huygens, see Koyré, *Newtonian Studies*, Aiton, *The Vortex Theory of Planetary Motions*, and Bertoloni Meli, *Equivalence and Priority*, 38–55. For Descartes's own treatment of gravity, which obviously predates Newton's *Principia* but which was salient for both Leibniz and Huygens, see *Principia Philosophiae*, in *Œuvres de Descartes*, vol. VIII-1: 3.40 and 4.20–7, and De Gandt, *Force and Geometry*, 118ff.; see also the discussion in Aiton, *Vortex Theory*, 55–6. As mentioned above, Newton's most extensive rebuttal of Descartes's views in *Principia Philosophiae* can be found in *De Gravitatione*, a text discussed in chs. 4 and 5.

to the complete characterization in question: in telling us, say, that the ether is the cause of gravity, we learn that the ether causes the various motions in question, and it does so in a way that is proportional to mass and distance. Therefore the import of Newton's claim that gravity causes the planetary orbits is carried over to the future account: the ether causes the free fall of bodies on earth and the planetary orbits. And the action of the ether can obviously be measured in precisely the way that gravity is measured, for the ether causes free fall and the planetary orbits proportional to mass and distance.

The ether that Newton describes in query 21 to the *Opticks* raises another issue. Some interpreters take the question of action at a distance to remain in this case, for Newton seems to describe the etherial particles as exerting repulsive force, and the latter may involve a short-range distant action (*Opticks*, 350).[51] But this seems to miss the fact that we can apply the distinctions in the *Principia* discussed above to the repulsive forces of the etherial particles just as we can to gravity itself. Thus the ether may be the physical seat – the "cause" – of gravity, and it may involve repulsive forces among its particles that, in turn, have their own physical seat, perhaps in some other medium. From the fact that the ether and its particles serve as the physical basis of gravity it does not follow that we know the physical characterization of the short-range repulsive forces that those particles may exert. Further work would be required. And of course we cannot determine *a priori* where this investigation of forces will lead us, nor where it will stop. Any invocation of forces by Newton must be understood within the context of his distinction between the mathematical and the physical treatment of force.[52]

If it is correct to read Newton's distinction between the two treatments of force as a technical one, in part because it helps to clarify the notion that gravity is a real force, then the correspondence between Leibniz and Clarke appears to be hampered by their joint acceptance of the interpretation that the distinction is non-technical. This results, in turn, in their failure to recognize the subtlety in Newton's contention that gravity causes various phenomena. Failing to recognize the subtlety leads each of them to misstep, in my view.

---

[51] Michael Friedman raised this issue with me in conversation. For discussion, see Westfall, *Force in Newton's Physics*, 395, and Hall, "Newton and the Absolutes," 282.

[52] See also the crucial passage from an unpublished addition to book III of the *Principia* quoted above in ch. 2.

As for Clarke's interpretation, Newton's agnosticism appears to be less strict than Clarke's expression of it. For Clarke, as we have seen, the *Principia*'s treatment of force ought to be understood as causally neutral, as a description of various motions that eschews any determination of their cause; the latter is left, apparently, until someone can present a physical treatment of gravity. This is, it seems to me, a perfectly reasonable interpretation if we take Newton's distinction as a non-technical one. But as we have seen, by gravity Newton does not intend to refer to certain tendencies to motion; rather, he means to signal the startling conclusion that the force that maintains the moon in its orbit is the very same one that we call "gravity" on earth. A description of certain tendencies to motion *à la* Clarke presumably does not commit one to the claim that the motions in question have the same cause.

Leibniz also appears to construe Newton's treatment of gravity in a non-technical fashion. In the course of providing an exact mathematical description of planetary motions, we are free to treat the motions as if they arose from an attraction between the heavenly bodies; this treatment must explicitly avoid any causal claims concerning the motions in question. Thus far Leibniz and Clarke concur. Since the *Principia*'s treatment of force is causally neutral on this view, any causal claim must be a component of a physical characterization of gravity, such as that provided by Leibniz's vortex theory. It is therefore natural for Leibniz to understand Newton's claim that gravity causes the planetary orbits as a component of a physical treatment of force in an ordinary or non-technical sense. But because Newton explicitly eschews any vortex-type theory, and because he explicitly admits that he does not know what the physical basis of gravity is, Leibniz thinks that he must retract the contention that gravity causes various phenomena. He then takes Newton's stubborn resistance to that retraction to entail that he is actually committed to the view, not that we do not know what physical basis gravity has, but that gravity is efficacious without any physical basis or medium, i.e. it is a property of material bodies in virtue of which they alter one another's states of motion across empty space. In other words, to insist that gravity itself causes motions in the absence of any attribution of a mechanism is, from Leibniz's point of view, to assert that there is action without a mechanism. And as we have seen, for Leibniz that means distant action. Clarke concurs with Leibniz's interpretation in the sense that he sees the denial of any causal commitment on Newton's part as his only viable means of escape from this pitfall. For his part, Newton does not follow the lead of his most prominent defender in that respect.

THE ONTOLOGY OF FORCE

For obvious reasons the discussion thus far has centered on Newton's conception of gravity, the force that serves as the centerpiece of book III of the *Principia*, and the attendant issue of action at a distance. But for some mechanists Newton's approach to force raises another issue, as I mention above. According to what I have called strict mechanism, all natural change arises solely from the surface actions of material bodies characterized by size, shape, motion, and solidity, so any contention that forces "exist" must be merely provisional. Forces must somehow reduce to the mechanist properties listed above. Just as the strict mechanist would deny that two material bodies can interact at a spatial distance, she would deny that when a body collides with another body it impresses a force on it, for there simply is nothing above and beyond bodies and their states of motion called a "force." To (strict) mechanist ears, the contention that forces exist would be understood as a claim that force should now be added to our previously austere ontology of size, shape, motion, and solidity.[53] And that is *verboten*.

Newton's response to the strict mechanist challenge is contained in embryo in his response to the dilemma discussed above. The *Principia*'s treatment of gravity, as we have seen, results in the law of universal gravitation, and that law, in turn, indicates that gravity is proportional to mass and is therefore not a mechanical cause (in Newton's technical sense). This allows Newton to make a move in his debate with the mechanists that they cannot foresee: he contends that we must distinguish local action from surface (or impact) action. The fact that gravity does not involve the latter does not entail that it does not involve the former: indeed Newton, like the mechanist, will insist that it must involve the former, on pain of unintelligibility. Similarly, Newton makes another move that the strict mechanist does not foresee: forces can exist without being part of our ontology in the strict mechanist sense. That is not because Newton denies that forces are part of our ontology in the mechanist sense – it is because he intends to address a separate question altogether. For him, as we have seen, forces exist because they are quantities that can be measured; and indeed, they can be measured by measuring other physical quantities that are perfectly uncontroversial, such as mass and distance. This settles certain issues that might

---

[53] Some interpreters of Newton, of course, think that Newton in fact adds force to our ontology in precisely this way. See Westfall, *Force in Newton's Physics*, 377–80, 384. Cf. also Cotes's discussion in his preface to the second edition of the *Principia*, 391–2, and Stein, "On Philosophy and Natural Philosophy," 195. I question this view in what follows.

be considered ontological – certainly it is crucial that forces can be measured, not least because there are physical features of the world that cannot be measured. Hence under certain conditions, the answer to the question – what is the ontology of force? – is simple: a quantity. But of course, this does not answer the mechanist question. What the mechanists could not foresee, and perhaps could never understand, is that Newton made the brilliant move of showing us how to measure certain fundamental quantities without understanding the *ontology* of those quantities in various senses of the term.

This general issue about force returns us to the debate with which I began. At this stage, it might seem that Cohen simply erred in claiming that Newton does not address the "existence of forces"; after all, Newton contends explicitly that gravity exists, as we have seen, so Westfall must surely be right. But Westfall's contention that Newton rejected the mechanical philosophy by approaching it on its own terms, insisting that we add gravity, or perhaps even force, to the list of primary qualities of bodies, is also problematic. For Newton never adds gravity to his list of the fundamental qualities of bodies; instead, he adds mass to the usual list of extension, mobility, and impenetrability (see ch. 4). It is easy to overlook the fact that one of the quantities Newton uses to measure force, namely mass – or the "quantity of matter" – is in fact the best candidate for the element that Newton adds to the ontology of the mechanist. For Newton is perfectly clear in book III of the *Principia* that he takes mass to be a "universal quality" of bodies akin to extension, impenetrability, and motion (*Principia*, 795–6). And if mass were not considered to be a property of bodies in the way that extension is, Newton's minimalist approach to the ontology of force would fail. Without mass, many forces in nature could not be measured, so Newton might not be in a position to state that a force is a quantity that can be measured by measuring other physical quantities. Moreover, Newton explicitly argues in book III that the weight of a body is proportional to its mass, and completely independent of its "form and texture." Hence bodies made of innumerable types of matter – Newton tested gold, silver, lead, glass, sand, salt, wood, water, and wheat (*Principia*, 807) – have weights that are proportional to mass, independent of their shapes or other properties. Thus it is mass, and mass alone, that is salient for the force of gravity, and not size, shape, motion, or impenetrability (*Principia*, 809). In that regard, by focusing on force, there is a sense in which the strict mechanist was looking in the wrong place: Newton does not add force, but rather mass, to our ontology. Whether the mechanists can accept Newtonian mass, however, is a further question (see ch. 4).

Perhaps mass should be added to the mechanist ontology, but does that move simply evade the issue at hand? Why should we not construe the claim that gravity exists as the claim that we should also add gravity to our ontology? Why should we not understand Newton's claim that gravity exists as the following claim: gravity is a property of all material bodies in our world. Newton thinks that we should not embrace this claim; the reason is contained in embryo in his response to the dilemma above. In contending that gravity exists, Newton is contending that a physical quantity with a particular measure exists – as we have seen, this quantity differs from magnetic force because they have distinct measures. Yet the knowledge that such a physical quantity exists does not amount to the knowledge that bodies have a certain property, even in our world. Such knowledge is dependent upon an account of gravity's physical cause – for instance, if gravity is a causal transmission through the medium of the ether, perhaps it is not a property of material bodies at all. Since we do not know, Westfall's interpretation grants us more knowledge than we actually have.

This reply, however, may simply serve to sharpen our understanding of Newton's problem. In my view, the claim that gravity exists amounts to the claim that a physical quantity exists, and yet I have also said that we do not know whether that quantity is a property of any object. But how can that be? Ordinarily, if one is measuring a quantity – a volume, a length, the distance from Rome to Paris, etc. – one has at least a decent fix on the ontology of the quantity being measured. By this I mean that if we measure some quantity, such as the length of a table, we can typically indicate whether it is a property of some substance, or instead a substance in its own right; we might even be able to indicate whether it is an intrinsic or a relational property, and so on. So Newton's view, on my construal of it, is bound to perplex his mechanist interlocutors and his contemporary readers: we are told that we are measuring a quantity when we measure a force such as gravity, but we have no idea whether we are measuring a property, a mode, or even a substance. If the physical treatment of force indicates the presence of an ether that serves as the medium for all gravitational interactions, then we might have measured the causal power of a substance by measuring gravity; on the other hand, the physical treatment of gravity might discover something entirely different.

Newton was perfectly well aware that his mechanist interlocutors would misinterpret his work, and I think the aspects of his early work in optics discussed above in ch. 2 may enable us to understand his position here. We can contend that light exists, and even measure some of its characteristics, such as its speed, without knowing whether it is a particle or a wave, or even whether it is a substance or a property of some kind of medium.[54] Newton took it upon himself to measure some of light's characteristics – for instance, the differential refrangibility of rays of light that emerge after ordinary sunlight passes through a prism – without knowing light's ontology, that is, without knowing whether it was wavelike or corpuscular. If light is a wave, is it really just a property of some medium, in the way that waves on the surface of a lake are certainly not substances, but rather properties of the water that constitutes the lake? Or is it a stream of particles? The fact that we can see rays of light, of course, may shortcircuit any skepticism about their existence, and this is disanalogous from the case of gravity. But the aim here is not to evade skepticism regarding the claim that gravity exists; instead, it is to clarify what that claim amounts to in the first place, within Newton's treatment of force. And the claim that light exists is not entirely dissimilar from Newton's claim about gravity, for in the case of light, we have no idea what it is that we are seeing, ontologically speaking. Newton never doubts that there is in fact some ontology underlying each thing, rays of light, and the force of gravity; but he denies that his existence claims in each case must be understood as claims that some property or some substance exists. That is why Newton's focus on measurement is absolutely central: in the absence of any traditional understanding of the ontology of our objects of study, we must rely on the measurement of various characteristics of the phenomena in question. If we lacked those measurements, we would be lost.

Since Descartes's physics analyzes bodily impacts through the lens of forces or powers, he may not be a strict mechanist, and his followers may therefore not react to Newton's physics as a strict mechanist would. Yet the contrast between Descartes's approach to force and Newton's own is nonetheless stark, and is especially salient here. Descartes mentions forces or powers in his third law of nature, which indicates what occurs when two bodies, with varying forces or powers, collide:

The third law of nature is this: when a moving body meets another, if its power [*vim*; *force*] of proceeding in a straight line is less than the resistance of the other body, it is deflected so that, while the quantity of motion is retained, only the direction changes; but if its power of proceeding is greater than the resistance of the other body, it carries that body along with it, and loses a quantity of motion equal to that which it imparts to the other. (*Principia Philosophiae*, VIII-1: 65)

---

[54] See especially Newton's letter to Hooke of 11 June 1672, sent through Oldenburg – *Correspondence*, vol. I: 174. For discussion, see Shapiro, *Fits, Passions and Paroxysms*, 21–4.

The Cartesian physicist knows *a priori* – i.e. independently of any empirical evidence, or of any particular physical theory – that forces or powers must be modes of corporeal substances. Since corporeal substances have extension as their essential property, forces must be modes of extension. In other words, forces must be identical with – or reducible without remainder to – the size, shape, and motion of bodies.[55] Hence the famous Cartesian vortex theory of planetary motion, or something akin to it that reduces gravity to mechanical interactions among geometrically characterized material bodies, is metaphysically required for Descartes.

The Cartesian response to Newton is clear. To Cartesian ears Newton's contention that gravity exists – or indeed that any force exists – can be understood in either of two ways. First, force can be construed as a term that refers to some quantity that reduces to some combination of the principal properties of objects – size, shape, and motion. Hence it might refer to the (size × motion) of a body. In that case, the contention that forces exist is philosophically innocuous. Second, if one insists that forces need not be reducible to mechanist properties, the Cartesian would resist the claim on purely *a priori* grounds. Newton reacts to this likely Cartesian construal by rejecting it: we can think of forces as measurable physical quantities without knowing how they fit into our basic ontology, which is precisely the situation with light.

In my view, then, we can split the difference between Cohen's and Westfall's interpretations as follows: on the one hand, Newton does address the existence of forces by claiming that "gravity really exists," so Cohen errs in that regard; but in making that claim, Newton is not claiming that we should add gravity to the strict mechanist – or the Cartesian – ontology, so Westfall errs in that regard. What unifies the different elements of Newton's overarching view, I think, is his subtle rejection of the terms of the debate within natural philosophy established by mechanists of various stripes. In responding to the special issue involving gravity, Newton rejects the mechanist presupposition that all local action must involve impact, but preserves the rejection of action at a distance, thereby opening up a conceptual possibility that his critics failed to envision. And in responding to the general issue, he denies that existence claims concerning physical quantities such as gravity must be construed within the terms of the strict mechanist ontology. If we can measure a physical quantity such as gravity, it exists just

---

[55] See the discussion in Garber, *Descartes's Metaphysical Physics*, 293–9. Unlike Newton, Descartes does not connect his notion of force, mentioned in the third law, to his notion of the quantity of motion, which for him is the scalar quantity, volume × speed.

as any physical quantity does, even if we remain ignorant of its ontological basis in the strict mechanist (or the Cartesian) sense.

The interpretation in this chapter answers questions about force that, in turn, raise two further questions, one about ontology, another about the relation between physics and metaphysics. First, I have suggested that if one is interested in ontology in the way that the mechanist might be – for instance, if one wants to know what the "primary qualities" of body are – then one might say that the *Principia* does not add force, or gravity, to our ontology, but rather mass. But what does this mean, and to what extent does Newton himself endorse this idea? This will be answered in the next chapter. Second, in my view, Newton takes the mechanists to err in thinking that we must avoid distant action by accepting only surface action; but Newton may also impose an *a priori* constraint on physics by rejecting action at a distance as inconceivable. Does he endorse such a constraint, and if so, how does it illuminate his view of the relationship between physical theory on the one hand, and metaphysics on the other? That question will be answered in the concluding chapter.

4

Matter and mechanism: contesting the mechanical philosophy, II

We have seen that gravity is a physical quantity, although Newton remains ignorant of how it fits into the ontology that most of his interlocutors would accept. He is ignorant of whether gravity is a property, a mode, or something else entirely. Many of Newton's interlocutors – Leibniz most prominent among them – balked at this profession of ignorance, arguing that at best, he tacitly treats gravity as a property of all material bodies. From their point of view, Newton had thereby revived an aspect of late Scholastic natural philosophy that the mechanists had attempted to bury, the doctrine of occult qualities. Newton and the editor of the *Principia*'s second edition, Roger Cotes, an astronomy professor at Trinity College Cambridge, attempted to rebut this charge, adding an extensive preface by Cotes in which he ridicules Newton's critics.[1] However, determining the proper response to this criticism broke the ranks of the Newtonians. Cotes insisted that gravity was a primary quality like any other, not occult but manifest. Newton resisted that response, avoiding the contention that gravity was any type of quality.

Newton's response to Leibniz's criticism required a delicate balancing act: he proclaimed explicitly that all bodies in the universe "gravitate" towards one another, even while denying that he took gravity to be a quality of bodies. As we saw briefly in the last chapter, we can understand this subtle view only if we recognize that mass, rather than gravity, is the new quality discovered in the *Principia*. This shifts the terms of Newton's engagement with the mechanical philosophy: although he may have avoided the invocation of occult qualities, just as he avoided action at a distance, he

---

[1] While Bentley was staying with Newton in March of 1713, Cotes sent them a letter in which he offered to let Newton write the editor's preface to the new, second, edition of the *Principia* (*Correspondence of Richard Bentley*, vol. II: 459–60). Bentley and Newton replied, on 12 March 1713, that Cotes ought to write the preface himself, and ought not to mention Leibniz by name, as it might be seen as "uncivil" – *Correspondence of Richard Bentley*, vol. II: 460–1. Cotes obliged them, although, as we will see, his views may diverge from Newton's own.

nonetheless ended up rejecting a powerful mechanist picture of the basic properties of material objects. The picture he substituted in its stead is complex enough to require this entire chapter to explicate.

Prima facie, the view that mass is a "universal quality" of bodies, or perhaps even that it is "essential" to matter, may strike us as considerably more straightforward than the claims about gravity's status discussed above. After all, in the case of mass we can dispense with the controversy involving action at a distance, and none of Newton's critics charged him with invoking an occult quality when he attributed mass to all bodies. Moreover, although mass is clearly crucial to Newton's physical theory, our ordinary conception of it is presumably not bound up with that theory. In that sense mass might seem to be a welcoming refuge for a Newtonian stung by the raging controversies regarding the *Principia*'s treatment of gravity. In tandem, whereas providing a "physical treatment" of Newtonian gravity is notoriously difficult – neither the vortex theory of Leibniz and some Cartesians, nor the ether theory of some Newtonians, gained much credence in the early eighteenth century – it would seem that a physical understanding of mass is easily articulated. Unlike gravity, mass is an ordinary intrinsic feature of material bodies, that is, a property that is independent of a body's relations and of its state of motion. But these appearances are deceiving. Although Newton's critics focused on what they characterized as Newton's invocation of the occult quality of gravity, as I discuss below, mass also raises problems from a mechanist point of view.

IS GRAVITY AN OCCULT QUALITY?

The contention that the Newtonian theory of gravity invokes an occult quality was not limited to Leibniz and his followers, or even to his mechanist critics. In an early passage in Berkeley's *De Motu*, for instance, we read:

While we support heavy bodies we feel in ourselves effort, fatigue, and discomfort. We perceive also in heavy bodies falling an accelerated motion towards the centre of the Earth; and that is all the senses tell us. By reason, however, we infer that there is some cause or principle of these phenomena, and that is popularly called *gravity*. But since the cause of the fall of heavy bodies is unseen and unknown, gravity in that usage cannot properly be styled a sensible quality. It is, therefore, an occult quality. But what an occult quality is, or how any quality can act or do anything, we can scarcely conceive – indeed, we cannot conceive. And so men would be better to let the occult quality go, and attend only to the sensible effects. (*De Motu*, §4, in *Philosophical Works*; Luce trans.)

So for Berkeley, we apparently have two options: we can take gravity to be the cause of certain phenomena – or more precisely, "gravity" is the name that we give to the cause of certain phenomena – in which case we must think of gravity as an occult quality of bodies; or we can eschew the view that gravity is a cause or a quality at all, attending only to sensible effects, viz. those effects that the Newtonians have associated with what they call "gravity." But clearly, in this context, to contend that gravity is a cause is to contend that gravity is a quality, and an occult one at that. Moreover, in the *Principles*, Berkeley describes the Newtonians as claiming that "attraction" is an essential property of all material bodies (*Principles*, 106). And then it is not much of a step to contend that this is an occult quality.

But Berkeley is a subtle critic, for he did not ascribe these views to Newton himself; instead, he provides an intriguing interpretation of the principal implication of Newton's treatment of force as follows:

Force, *gravity, attraction*, and terms of this sort are useful for reasonings and reckonings about motion and bodies in motion, but not for understanding the simple nature of motion itself or for indicating so many distinct qualities. As for attraction, it was certainly introduced by Newton, not as a true, physical quality, but only as a mathematical hypothesis. (*De Motu*, §17, *ibid.*)

So Berkeley envisions at least two ways of avoiding the invocation of an occult quality: first, one can eschew the term gravity, speaking entirely of sensible effects, especially various motions. Second, one can employ the term *gravity*, but only as a component of a mathematical hypothesis that allows for the prediction of the motions of bodies, and not as the name of a physical quality. In the one case we eschew the term, and in the other it presumably does not refer. In either case the qualities of bodies will not include "gravity," and the mathematical notion of gravity will enable the prediction of changes in these qualities, especially changes of bodily motions and spatial positions.

Berkeley's reasonably subtle analysis can be read as foisting a choice on Newton. In order to avoid treating gravity as an occult quality, one can either eschew the term altogether, attending only to sensible effects, such as motions; or one can employ it as a non-referring term within a theory, one that allows for the prediction of sensible effects. One way of understanding Clarke's response to Leibniz is to see him as endorsing this formulation of the difficulty facing the Newtonian – Clarke then argues that either choice is acceptable. But if my analysis in ch. 3 is correct, Newton himself rejects this formulation, arguing that gravity actually exists – hence the term in his theory refers – but is nonetheless not an occult quality.

Roger Cotes may provide the strongest Newtonian rejection of the Berkeley formulation, and his ultimate break from Newton is especially illuminating. In his preface to the second edition of the *Principia*, before Cotes responds to Leibniz's criticisms, he accuses other philosophers of invoking occult qualities. He writes:

Those who have undertaken the study of natural science can be divided into roughly three classes. There have been those who have endowed the individual species of things with specific occult qualities, on which – they have then alleged – the operations of individual bodies depend in some unknown way. The whole of Scholastic doctrine derived from Aristotle and the Peripatetics is based on this. Although they affirm that individual effects arise from the specific natures of bodies, they do not tell us the causes of those natures, and therefore they tell us nothing. And since they are wholly concerned with the names of things rather than with the things themselves, they must be regarded as inventors of what might be called philosophical jargon, rather than as teachers of philosophy. (*Principia*, 385)[2]

The generality of Cotes's remarks is typical in this period. This passage is apt because it illustrates a common problem with criticisms that reference occult qualities, viz. they are sufficiently polemical to lack a clear statement of the criticism.

For our purposes here we can treat a quality as occult if it meets three conditions.[3] First, the quality is understood to cause effects in one or more other objects. Second, the quality is exhaustively characterized by these effects. And third, the quality is said to be distinct from, and irreducible to, the primary qualities of its bearer. Two criticisms are typically leveled against a quality meeting this threefold definition: (1) that it is unintelligible or, more specifically, that one cannot have an idea of it and therefore, according to some, that one cannot have a coherent thought about it, but can only talk about it; and (2) that it fails to be explanatory. Cotes clearly has (1) in mind when mentioning the mere "jargon" of his predecessors. With respect to (2): it may not be the case that a quality that meets the three conditions fails to explain just any natural phenomenon. Rather, (2) should

---

[2] The second class of natural philosophers appear to be the Cartesians, and the third are the Newtonians (*Principia*, 385–6). Of course, Cotes's preface is intended to be polemical, and therefore may present Newton's Cartesian and Leibnizian opponents in an unflattering light.

[3] My usage here reflects what I take to be the most salient objections to Newton's theory in the late seventeenth and early eighteenth centuries, especially in Leibniz, who raised the objection in his correspondence with Clarke (L 5: 113, *Die philosophischen Schriften*, vol. VII: 417). For a fascinating discussion of the shifting semantics of *occult quality* from the first half of the seventeenth century to the second half, see Hutchison, "What Happened to Occult Qualities in the Scientific Revolution?", 250ff.

really be taken to mean: the quality cannot explain the sensible effects through which it is said to be exhaustively characterized. This captures a problem with the ubiquitous dormative virtue case.

These three conditions help to illuminate significant differences among Cotes's various uses of *occult* when criticizing Newton's predecessors and defending the *Principia*. Just after concluding that gravity is in fact a universal quality of matter – a view I tackle below – Cotes writes:

> I can hear some people disagreeing with this conclusion and muttering something or other about occult qualities. They are always prattling on and on to the effect that gravity is something occult, and that occult causes are to be banished completely from philosophy. But it is easy to answer them: occult causes are not those causes whose existence is very clearly demonstrated by observations, but only those whose existence is occult, imagined, and not yet proved. Therefore gravity is not an occult cause of celestial motions, since it has been shown from phenomena that this force really exists. Rather, occult causes are the refuge of those who assign the governing of these motions to some sort of vortices of a certain matter utterly fictitious and completely imperceptible to the senses.

> But will gravity be called an occult cause and be cast out of natural philosophy on the grounds that the cause of gravity itself is occult and not yet found? (*Principia*, 392)

From Cotes's point of view, gravity itself is not occult, only its cause is.[4] This indicates, first of all, an ambiguity in the term occult, for the term can mean at least one of two things. In what I will call its non-technical sense – where it does not meet the threefold definition above – saying that the cause of gravity is occult means simply that the cause is hidden or unknown. But in the technical sense sketched above, the claim means that the cause of gravity is an unintelligible power of bodies. Cartesian and Leibnizian readers must have been outraged that Cotes characterized their vortex theories of planetary motion as involving "occult causes," for their argument was precisely that vortices enable us to meet the explanatory strictures of the mechanical philosophy, attributing planetary motions to an eminently

---

[4] It may sound odd to discuss gravity's "cause" – for Cotes and Newton, this refers to gravity's physical basis, as I note in ch. 3. For instance, in his second letter to Bentley, Newton writes: "You sometimes speak of gravity as essential and inherent to matter. Pray do not ascribe that notion to me; for the cause of gravity is what I do not pretend to know" (Newton to Bentley: January 17 1692/3, in *Philosophical Writings*, 100). This mirrors the view in the General Scholium to the *Principia*, added to the second edition of the text (1713). Thus to characterize gravity's physical basis – here, to claim that it inheres in matter *per se* – is to characterize its "cause." The "cause" of gravity might be the ether or some other medium, according to Newton; it might be the essence of matter itself, independent of any medium, according to others.

intelligible impact mechanism.[5] Perhaps Cotes means that vortices are occult in the ordinary sense, for he takes them to be imperceptible and therefore hidden. He presumably does not mean that they involve unintelligible powers.

The same may be true of what Cotes calls gravity's cause. If Cotes has the ordinary, rather than the technical, meaning in mind, then gravity's cause is something that he could have hoped to discover, even if he did not know it in 1713. This is a reasonable point for Cotes to make, for it seems to me as clear as anything that in each edition of the *Principia*, Newton thinks that it is perfectly sensible to search for a cause of gravity, which is to say, in the least, that he certainly thinks that gravity does have some cause.[6] If Cotes has the technical sense in mind, however, then gravity's cause is not something that we could hope to discover, not because it is hidden or microscopic, but because it is unintelligible and therefore not the sort of thing that could be discovered. So Cotes's view has two aspects: first, gravity is not occult because it is a manifest quality; and second, gravity's cause is not occult if by that we mean more than the claim that its cause is hidden.

But in what sense is gravity manifest? In answering this question, Cotes fearlessly wades into deep waters by rejecting the Berkeleyan formulation with a stark ontological claim. He writes:

The extension, mobility, and impenetrability of bodies are known only through experiments; it is in exactly the same way that the gravity of bodies is known. All bodies for which we have observations are extended and mobile and impenetrable; and from this we conclude that all bodies universally are extended and mobile and impenetrable, even those for which we do not have observations. Thus all bodies for which we have observations are heavy; and from this we conclude that all bodies universally are heavy, even those for which we do not have observations. If anyone were to say that the bodies of the fixed stars are not heavy, since their gravity has not yet been observed, then by the same argument one would be able to say that they are neither extended nor mobile nor impenetrable, since these properties of the fixed stars have not yet been observed. Need I go on? Among the primary qualities of all bodies universally, either gravity will have a place, or extension, mobility, and impenetrability will not. (*Principia*, 392)[7]

---

[5] See, for instance, Leibniz's presentation of a vortex theory of the planetary orbits in his *Tentamen* of 1689 (written after the publication of the *Principia*) in *Die mathematische Schriften*, vol. VI; and in English translation in Bertoloni Meli, *Equivalence and Priority*. See also Leibniz's letter to Newton of 1693, *Philosophical Writings*, 106–7.

[6] I discuss this issue, and related matters, in some depth in ch. 3.

[7] Cotes writes here of the *primarias qualitates corporum* – see *Principia Mathematica*, vol. I: 27. Apparently, Cotes originally wrote that gravity was "essential" to matter, but Samuel Clarke asked him to change his formulation when Cotes submitted a draft of his preface for Clarke's approval. If

Although this claim proved deeply controversial, it was nonetheless rather natural for Cotes to make it, for Newton himself employs the third of his famous *regulae philosophandi*, to which Cotes here alludes, to prove that gravity acts universally. Newton's third rule reads as follows: "Those qualities of bodies that cannot be intended and remitted and that belong to all bodies on which experiments can be made should be taken as qualities of all bodies universally" (*Principia*, 795–6). In every other case Newton employs rule three to buttress an inference of this type: from the fact that some quality is found within the reach of our experiments we infer, given certain background conditions, that it is a quality of all bodies.[8] And so Cotes reasonably concludes that gravity, too, is a quality of bodies universally, thereby attempting to block two criticisms: one, that gravity is a quality of bodies, but not of bodies universally (this involves the scope of the inference we may make from our empirical evidence); and two, that gravity must be considered an occult quality if it is a universal quality, a view he rebuts in the paragraph quoted above, which in fact follows the paragraph I have just quoted.

Cotes and Newton corresponded extensively while preparing the second edition of the *Principia*, but Newton resisted Cotes's inference that gravity was a primary, or a universal, quality (not that it is a universal force). Before I explore his resistance, however, one point merits discussion here. Even if Newton had followed Cotes's lead, he would have had grounds for denying that he treated gravity as an occult quality, for it failed to meet the second condition in the technical sense of *occult*. As Newton understands matters,

Clarke sent Cotes a letter, it has not survived; he may have used an intermediary, a "Dr. Cannon," mentioned by Cotes. Cotes's reply to Clarke reads in part as follows:

Sir – I return to you my thanks for your corrections of the preface, & particularly for your advice in relation to that place where I seemed to assert gravity to be essential to bodies. I am fully of your mind that it would have furnished matter for caviling and therefore I struck it out immediately upon Dr. Cannon's mentioning your objection to me, & so it never was printed. My design in that passage was not to assert gravity to be essential to matter, but rather to assert that we are ignorant of the essential properties of matter & that in respect to our knowledge gravity may possibly lay as fair a claim to that title as the other propertys which I mentioned. For I understand by essential propertys such propertys without which no other belonging to the same substance can exist: and I would not undertake to prove that it were impossible for any of the other propertys of bodies to exist without even extension. (*Correspondence of Cotes and Newton*, 158; also available in *Correspondence*, vol. V: 412–13)

For discussion, see Koyré, *Newtonian Studies*, 281. In his famous "Newtonian" notes to Rohault's treatise, *System of natural philosophy*, Clarke himself indicates that Newton has shown gravity to be a "property of matter" [*materiae proprietatem*], although not an essential one – *System of natural philosophy*, vol. II: 96 n.; for the original Latin, see Clarke's *Annotata* to Rohault, *Physica*, 80 (which is the note to part II, ch. 28, article 13). For a discussion of Clarke's notes that focuses on the philosophical implications of the theory of universal gravity, see Metzger, *Attraction universelle*, part 3: 115–17.

[8] See the perceptive account in Mandelbaum, *Philosophy, Science and Sense Perception*, 61–117.

it is essential to gravitational interactions that the force of gravity between any two objects is "as the masses" (cf. Newton, *Treatise on the System of the World*, 33–5), so if gravity were a quality it could not be exhaustively characterized by the effects it generates. This is an easily missed point, one that may have significant consequences for certain eighteenth-century, largely empiricist, construals of Newton's theory. Suppose one wanted to construe gravity as a power borne by the sun to "attract" the earth. The problem is that a complete description of the motions of the sun and the earth through space, including an absolutely precise diachronic description of the earth's solar orbit, would fail exhaustively to characterize the gravitational interaction of the earth and the sun. This would be true even if they were otherwise lonely bodies. So the claim that the sun "attracts" the earth cannot be translated into the claim that the sun has a "power" to produce various observed earthly motions through the solar system, for any power the sun might bear is proportional to an intrinsic feature of the earth, and the latter cannot be explicated in terms of the earth's motion. This highlights the significance of Newton's view – which I discuss below – that mass is an intrinsic feature of a body, one independent of its state of motion and of its gravitational interactions with other bodies.[9]

This underscores the importance of Newton's reluctance to endorse Cotes's contention that gravity is a primary quality. For it seems that Newton could have had his cake and eaten it, too – he could have claimed that gravity was a primary quality even while rebutting the charge that this claim committed him to treating gravity as occult. Specifically, he could have claimed that gravity was harmlessly occult, which is to say, perceptually hidden, rather than occult in the technical sense, which is to say, an unintelligible power. Instead, Newton carefully eschewed Cotes's view, and he did so in the second edition of the text, which Cotes edited.[10] The question is, why?

Recall the third of the *regulae philosophandi*: "Those qualities of bodies that cannot be intended and remitted and that belong to all bodies on which experiments can be made should be taken as qualities of all bodies universally" (*Principia*, 795–6). Newton discusses this rule and its implications as follows:

---

[9] If we take mass to be a primary quality of bodies, then gravity *à la* Newton fails to meet the third criterion as well. Any power the sun might have to "attract" the earth would not be independent of, or distinct from, the sun's primary qualities, since it would be proportional to the sun's mass.

[10] Although Newton and Cotes corresponded extensively about the second edition of the *Principia* (see ch. 6 for some details), Newton evidently did not see Cotes's preface before it was published (Westfall, *Never at Rest*, 749).

The extension, hardness, impenetrability, mobility, and force of inertia of the whole arise from the extension, hardness, impenetrability, mobility, and force of inertia of each of the parts; and thus we conclude that every one of the least parts of all bodies is extended, hard, impenetrable, movable, and endowed with a force of inertia. And this is the foundation of all natural philosophy . . . Finally, if it is universally established by experiments and astronomical observations that all bodies on or near the earth gravitate toward the earth, and do so in proportion to the quantity of matter in each body, and that the moon gravitates toward the earth in proportion to the quantity of its matter, and that our sea in turn gravitates toward the moon, and that all planets gravitate toward one another, and that there is a similar gravity of comets toward the Sun, it will have to be concluded by this third rule that all bodies gravitate toward one another . . . Yet I am by no means affirming that gravity is essential to bodies. By inherent force I mean only the force of inertia. This is immutable. Gravity is diminished as bodies recede from the earth. (*Principia*, 796)

So Newton assiduously avoids listing gravity along with "extension, impene-trability, mobility, and force of inertia"[11] as what he calls a universal quality of bodies. By "force of inertia," Newton means mass, as he indicates in the third definition at the opening of the *Principia* (404–5). It seems that Cotes rushed in where Newton feared to tread (which is not to say that Cotes is a fool).

Two questions now face us. First, am I right that Newton eschews the view that gravity is a universal quality? After all, the passage just quoted is not entirely clear on this point. And second, if this interpretation is correct, why does Newton endorse this position?

Regarding the first question, Newton certainly could have been more careful in his discussion of rule three, for it can look as if he infers that a quality within the reach of our experiments that is not subject to being increased or diminished is in fact a universal quality. But actually, it is just the form of the inference regarding gravity that licenses its placement within the discussion of the third rule, for by Newton's own lights – as he himself says in this very passage – gravity in fact diminishes with distance. This is unlike mass, extension, impenetrability, and mobility, which do not increase or diminish with a body's distance from any other body, or indeed with any change of relations between the bearing body and any other body. So Newton explicitly notes that gravity fails to meet one of the criteria

---

[11] Given Cotes's discussion in the preface, which explicitly mentions "primary qualities," and given the tradition of thinking about primary and secondary qualities in the late seventeenth century, inter-preters sometimes slip by taking Newton to be discussing a criterion for a quality's being primary (see, for instance, Harré's discussion in *Matter and Method*, 106–7). In fact, he discusses the question of which qualities are "universal" and which are "essential to matter." Note: since it is not pertinent here, I bracket "hardness" (inelasticity).

enabling the inference in question; therefore, that cannot be Newton's inference. Instead, Newton claims that if the bodies within the reach of our experiments "gravitate" toward one another, we can conclude that all bodies do so. So the empirical premise in the inference is distinct from the corresponding premise in the other cases. In this case, we do not consider a quality of bodies, but rather a type of interaction.

The importance of the claim that gravity is a universal type of interaction is easily missed. It distinguishes Newton's view of gravity from his view, for instance, of magnetism, since not all bodies are subject to magnetic force. It is crucial that if we take mass to be an essential property of material bodies – which Newton explicitly does, and which I discuss in much greater depth below – then it follows from Newton's theory that all bodies in the universe bear gravitational interactions (and indeed, with all other bodies). This is not the case for magnetism and other forces. So one can emphasize that gravity is a type of interaction rather than a quality, but to indicate its placement within the discussion of rule three, one might emphasize instead that it is a universal type of interaction, rather than a particular one.

This leaves us with the second question from above: why does Newton begin his inference with a premise concerning a type of interaction, rather than one concerning a quality?[12] This hinges on Newton's conception of what we are licensed to conclude about gravity from the treatment of force in the *Principia* (see ch. 3). Long before Newton had ever been accused of invoking an occult quality, he gave us a fascinating hint as to how he himself would attempt to rebut that charge later in his life. Already in 1687, in his preface to the first edition of the *Principia*, Newton contrasts investigating nature by invoking occult qualities with his own approach:

Since the ancients (according to Pappus) considered *mechanics* to be of the greatest importance in the investigation of nature and science and since the moderns – rejecting substantial forms and occult qualities – have undertaken to reduce the phenomena of nature to mathematical laws, it has seemed best in this treatise to concentrate on *mathematics* as it relates to natural philosophy. (*Principia*, 381)[13]

---

[12] More than any other contemporary interpreter, Stein has emphasized this aspect of Newton's thought – see, for instance, his remarks in "Newton's Metaphysics," 287–9. Stein quotes an extensive discussion of Newton's in his *De Mundi Systemate* (available in translation as *A Treatise of the System of the World*) at 305–6 n. 71.

[13] In discussing what he himself calls "occult qualities," Newton consistently – from 1687 to 1727 – contrasts these qualities with what he calls laws of nature or principles, which latter would be the focus of the mathematical treatment of force. The *Oxford English Dictionary* entry for "principle" includes the following two sub-entries. First, I.2: "That from which something takes its rise, originates, or is derived; a source; the root (of a word)." And second, I.3, in generalized sense: "A fundamental source from which

This passage contains a clue regarding the proper interpretation of the claim that Newton's theory treats gravity as an occult quality of material objects – the key lies in finding the proper interpretation of Newton's treatment of force in the *Principia*.[14]

As we saw in ch. 3, from Newton's perspective the contention that "gravity really exists" should be understood to mean that a physical quantity measured by the law of gravitation exists, and not that material objects bear some kind of property. For Newton, Cotes fails to recognize the import of our ignorance of gravity's "cause" – until we discover gravity's "physical cause," we are not in a position to say that gravity is a property of material bodies, for it may be a property of the ether or of some other medium. That is why in rule three, Newton treated gravity as a form of interaction, rather than as a quality, for we can measure only the physical quantity bound up with interactions among material bodies.[15] Newton's formulation in rule three, then, reflects the ignorance of gravity's cause that Newton professes. More importantly, Newton suggests that gravity is a type of interaction, rather than a quality, because it decreases with an increase in spatial separation, so it is sensitive to the spatial relations of two or more bodies. In that sense we might characterize it as a kind of spatial interaction between bodies.

Even more remarkably, there are grounds for thinking that Newton also conceives of gravity as a type of interaction, rather than a quality, in the case of a lonely massive body. It may sound odd to conceive of a single body as bearing any kind of interaction, but Newton has the resources to clarify this idea. In definition eight (*Principia*, 407–8) he distinguishes what he calls the "motive, accelerative, and absolute" measures of centripetal force. Howard Stein has suggested that in this definition Newton articulates something akin to the idea of a field of force.[16] One reason is that the "accelerative"

---

something proceeds; a primary element, force, or law which produces or determines particular results; the ultimate basis upon which the existence of something depends; cause, in the widest sense." Each sense will be relevant in what follows.

[14] As will become clear in what follows, this chapter presupposes the interpretation of Newton's distinction between the mathematical and the physical treatment of force outlined in depth in ch. 3.

[15] This might indicate an instance in which Locke's famous discussion of God's "superaddition" of gravity to matter does not quite reflect Newton's own understanding of the issue. For Locke seems to presuppose that the question facing the mechanist – and *mutatis mutandis*, any philosopher – is how gravity could possibly be a quality or a property of material bodies. In response to Newton's theory he concludes that although we cannot understand how matter can gravitate towards other matter, nonetheless God must have superadded gravity to matter. Newton clearly does not think that this is the question facing us; that is, the puzzling aspects of gravitational interactions do not include Locke's question of how gravity could be a property of material objects. See Locke's second reply to Stillingfleet in *Works*, vol. III: 460–2, 467. See note 64 below.

[16] See especially "On the Notion of Field," 266–8, and "Newton's Metaphysics," 286–7.

force may be referred "to the place of the body as a certain efficacy diffused from the center through each of the surrounding places in order to move the bodies that are in those places" (*Principia*, 407). This indicates that we can measure the centripetal force generated by a lonely body through the effect it would have on any massive body placed at any distance from the lonely body's center.[17]

But Newton's view is even more complex than this discussion suggests. For his treatment of gravity's status in rule three shifts the discussion away from a general question concerning the qualities of matter to the specific question of whether a quality characterizes the essence of matter. Newton seems to think that the qualities listed in rule three – extension, hardness [inelasticity], impenetrability, mobility, and mass – are essential to material bodies, that is, all material bodies bear these properties, and they cannot fail to bear them, on pain of failing to be material. If the properties are essential, in turn, they presumably meet the well-known lonely corpuscle criterion – even a lonely material body, or particle, would be characterized by the properties Newton lists. When this is the question on the table, Newton's ignorance of gravity's cause does not exhaust his reasons for denying that gravity is an essential quality – he wants to deny that gravity is essential to matter on distinct grounds, independent of his ignorance. If gravity were an essential property of material bodies, then two lonely bodies would bear the property, even in the absence of a medium and even if they were spatially separated. Thus if gravity were essential to them, it could not be diminished or removed through spatial separation, or through the disappearance of any medium between them. Hence the view that gravity is essential to matter appears to entail the claim that material bodies can act at a distance on one another.[18]

Newton is perfectly clear about this entailment whenever he discusses action at a distance, or the question of gravity's status. Consider, again, his famous caveat to Bentley, which he sent in late February of 1693:

---

[17] Thanks to Eric Schliesser and George Smith for discussing this issue with me.

[18] Although the claim that gravity is essential to matter entails that material bodies can act at a distance, the latter claim does not entail the former claim. Newton apparently never discusses this nuance, but it seems possible for bodies to act on one another at a distance in a world in which gravity is an accidental property of matter. Imagine a world in which there are two material bodies that exist at some spatial distance from one another, and that interact gravitationally. It might be the case that their gravitational interaction diminished as their distance increased – for Newton, this means that gravity is not essential to the bodies, as he notes in rule three. In that case, there would be a world in which bodies acted at a distance, but in which gravity was not an essential property of the bodies. Thanks to Eric Schliesser for discussion of this issue.

It is inconceivable that inanimate brute matter should, without the mediation of something else, which is not material, operate upon and affect other matter without mutual contact, as it must be, if gravitation in the sense of Epicurus, be essential and inherent in it. And this is one reason why I desired you would not ascribe innate gravity to me. That gravity should be innate, inherent and essential to matter, so that one body may act upon another at a distance through a vacuum without the mediation of anything else, by and through which their action and force may be conveyed from one to another, is to me so great an absurdity, that I believe no man who has in philosophical matters a competent faculty of thinking can ever fall into it. Gravity must be caused by an agent acting constantly according to certain laws; but whether this agent be material or immaterial, I have left to the consideration of my readers. (*Philosophical Writings*, 102)[19]

Newton embraced this view throughout his career. For instance, in a letter to Newton on 15 July 1719, Pierre Varignon writes that he cannot see "how anyone can suppose that you [i.e. Newton] consider gravity to be among the essential attributes of bodies," given what Newton says in definition eight of book I of the *Principia*, and in query 31 in the *Opticks*. On 29 September 1719, Newton replied as follows to Varignon's comment:

Besides the words you have cited, there are others in the Scholium to Proposition 69, Book I, *Principia*, from which it is clear that I have not in the past claimed that gravity is either essential to bodies or without the action of a medium. What I have added in the Queries at the end of the *Opticks* in the second edition [the second Latin edition] regarding the cause of refraction, reflection and gravity, I touched on more lightly in the first edition, Proposition 12, Part III, Book II. And there too it appears that I have neither demanded an absolute vacuum nor denied an etherial elastic medium diffused through all the heavens, by the action of which bodies can be deflected from their rectilinear motion. (*Correspondence*, vol. VII: 63)

So from Newton's point of view, we must embrace two distinct positions regarding gravity's status: first, we must reflect our ignorance of gravity's cause by eschewing Cotes's position that gravity should be counted as a quality of all material bodies; and second, in order to avoid the invocation of distant action, we must avoid the view that gravity is an essential property of material bodies.

Newton's two views, in turn, jointly suggest a future path of empirical investigation regarding gravity's cause: the impossibility of action at a distance already indicates that we could not discover that gravity was an essential property of material objects. We will have to discover some kind of medium such as an ether, that in some sense accounts for the gravitational

---

[19] We find the same point in Newton's letter to Bentley of 17 January 1693 (*Philosophical Writings*, 100).

interactions among material bodies, consistent with the law of universal gravitation. Since we are ignorant of such a medium, and ignorant of its characteristics, we cannot say whether gravity is some kind of non-essential property of material bodies. Perhaps it is a non-essential, relational property of bodies; but then again, it may be best to conceive of it as a property of the ether (perhaps a kind of relational property).

The very idea that the ether might serve as the medium for gravitational interactions raises several complex issues within the Newtonian context. Previous thinkers, such as Descartes, envisioned a material and mechanical ether as the medium for various kinds of causal propagation and interaction.[20] For Descartes the ether would differ from other kinds of matter only in the size, shape, and motion of its parts, and perhaps in their overall configuration. Hence the ether would effect natural change only by having its particles impact on other bodies, or other particles, and those, in turn, would then impact upon other bodies and particles. In that sense, the ether would be metaphysically indistinct from ordinary bodies; and, of course, it would follow Descartes's three laws of nature, so it would also be physically indistinct from ordinary bodies. Newton rejects the Cartesian ether because its particles would impact on the surfaces of the comets traversing the solar system, thereby impeding their motion (which is not observed astronomically).

Unlike the Cartesian ether, Newton's ether[21] – mentioned in various published writings, including query 21 to the *Opticks* (added to the 1717 edition) and the General Scholium to the *Principia* (1713) – would have to be non-mechanical in order to cohere with Newton's gravitational theory (as discussed in ch. 3). It would interact with the masses of bodies, flowing through them rather than impacting solely on their surfaces. The ether would presumably have a negligible mass, otherwise it would exert a

---

[20] See the discussion in Cantor and Hodge's introduction to *Conceptions of Ether*, 12ff., and in Aiton, *The Vortex Theory*.

[21] Newton's own shifting conceptions of the ether reflect the very issues discussed here, so his pre- and post-*Principia* discussions of the properties of the ether differ fundamentally from one another. In his famous 1679 letter to Boyle, Newton postulates a quasi-Cartesian – that is, a mechanical and material – ether whose particles impact upon bodies, thereby causing them to fall freely toward the earth's surface. Once he obtains the law of universal gravitation in the first (1687) edition of the *Principia*, however, Newton may no longer endorse this conception of gravity's cause – although some argue that he still leaves it open whether an ether operating on an impact model could in fact account for gravity. But crucially, with the publication of the second edition of the text in 1713, Newton's view shifts: he now explicitly denies that gravity is mechanical in an important sense, as we saw in ch. 3, and therefore requires a non-mechanical, ubiquitous ether in order to render all gravitational interactions merely local and proportional to mass. Thanks to Michael Friedman for discussion of this issue.

gravitational pull on other bodies, which would complicate any calculation in book III of the *Principia* in which we treat the planets as exchanging momentum with the sun directly, i.e. with no massive bodies in between (see ch. 6).[22] Newton presumably cannot contend that the ether is non-negligibly massive but does not exert a gravitational pull on other bodies, for then he would be forced to concede that gravity was not proportional to mass. That, in turn, would obviously be disastrous for the *Principia*, for Newton argues in corollary two to proposition 6 of book III that in fact gravity is independent of the form and texture of any body, depending only on mass. So any Newtonian ether would be non-mechanical and bear negligible mass, enabling us to ignore its mass in calculations of momentum exchange. The spatial ubiquity of the ether would guarantee that all gravitational interaction involved local action, so the ether could underwrite the local, non-mechanical, and spatially ubiquitous actions of gravity.

At other moments in his career Newton contemplated another spatially ubiquitous medium, something distinct from the ether just described. For instance, in his unpublished response to Leibniz's letter published in the *Memoirs of Literature* in 1712, he contends that

if any man should say that bodies attract one another by a power whose cause is unknown to us, or by a power seated in the frame of nature by the will of God, or by a power seated in a substance in which bodies move and float without resistance and which has therefore no *vis inertiae* [mass] but acts by other laws than those that are mechanical: I know not why he should be said to introduce miracles and occult qualities and fictions into the world. (*Philosophical Writings*, 116)

So the substance envisioned here would be distinct from the ether in being non-massive, and would therefore not follow the laws of motion. This of course raises questions of its own that we cannot delve into here.

Leaving aside the complexities of the ether, or other medium, Newton seems to be on firm ground in two respects when tackling the issues Cotes raises. We can understand his motivation for avoiding Cotes's contention that gravity is a quality of bodies, and for his denial that it is an essential property of bodies. Moreover, as we have seen, gravity's proportionality to mass allows Newton to avoid the claim that he treats gravity as an occult quality in the technical sense outlined above, even without endorsing Cotes's view. That is, Newton can deny both that gravity is occult and

---

[22] See the discussion in the Scholium to book II, proposition 40, where Newton criticizes the Cartesian vortex approach to celestial motion (*Principia*, 761), indicating that planets and comets exhibit no sensible diminution of their motions. Any celestial matter filling the heavens would show resistance to planetary motion in proportion to the mass of the matter in question.

that it is essential to matter even while remaining agnostic on its ultimate physical basis. But as Newton's discussion of his third rule indicates, mass plays a crucial role here that I have de-emphasized so far: it serves as a contrast class to gravity for, unlike gravity, it is both a universal quality of bodies and "essential" to matter. In the next section I want to explore this point in more depth.

DESCARTES AND NEWTON ON MATTER'S ESSENCE

Like many natural philosophers in the mid and late seventeenth century, Newton rejects what he took to be the principal means of explaining natural phenomena employed by late Scholastic, or neo-Aristotelian, thinkers. In the preface to the first edition of the *Principia*, Newton writes that "the moderns – rejecting substantial forms and occult qualities – have undertaken to reduce the phenomena of nature to mathematical laws" (*Principia*, 381). He clearly numbers himself among the moderns. But Newton explicitly eschews the Cartesian route to removing the last vestiges of Scholasticism from natural philosophy. Indeed, he rejects the Cartesian conception of matter, and its attendant assumptions and entailments regarding physics and its relation to metaphysics, on both metaphysical and epistemic grounds. I tackle the former in this section, and the latter in the following section below. Many of Newton's definitions in the *Principia* reflect his dissatisfaction with fundamental notions in Descartes's *Principia Philosophiae*, so we can highlight the novelty and complexity of Newton's "quantity of matter" (or mass) by comparing it with its well-known Cartesian ancestor, the *quantitas materiae*.[23]

From Descartes's point of view, of course, the late Scholastic conception of form and matter – one according to which bodies have innate tendencies to interact causally with other bodies – ought to be replaced with a much more austere metaphysical picture: bodies are characterized solely by size, shape, and motion, and all changes that they undergo are the result of impacts among them or their parts. Hence rocks do not have a property called "heaviness," they do not have a form that explicates their tendency to fall to the ground; instead, their descent to the earth's surface must be explained through the impacts of various other bodies or particles on their

---

[23] For discussions of Descartes's concept of the *quantitas materiae*, see Clarke, *Occult Powers and Hypotheses*, 71ff.; Blackwell, "Descartes' Concept of Matter," especially the note to p. 69; Jammer, *Concepts of Mass in Classical and Modern Physics*, 71ff.; and, the comment of the translators, V. R. and R. P. Miller, in Descartes, *Principles of Philosophy*, 152 n. 120. Cf. also Suppes's helpful discussion in "Descartes and the Problem of Action at a Distance," 146 n. 5.

rocky surfaces.[24] There are therefore no "specific" differences between types of matter – all material bodies have the same basic properties, so one body can differ from another only in ways that are compatible with this austere ontology. Descartes famously rejects the Aristotelian notion of natural place, arguing that space and bodies in the celestial and terrestrial realms should be understood through the same mechanist principles.

Part of Descartes's austere metaphysical picture can be characterized as follows. First, extension is the principal attribute of body, so that all the basic properties of any given material body – its size, shape, and motion – are aspects of its extension. Second, all causation within nature must be consistent with this austere ontology. Third, Descartes denies the distinction between body and space – we can certainly fail to perceive any body in a given space, but space cannot actually be empty of body, for extension is a property that must inhere in some substance. Hence Descartes rejects the vacuum, arguing that our world consists of a plenum; this entails, in turn, that all motion must involve a change in relations among bodies, for any change in spatial position just is a change in object relations within the plenum. In that regard Descartes's conception of body already entails part of our understanding of causation, of space, and of motion, and these are obviously crucial to any work in physics.[25]

The view that extension is the essence of matter, and the principal attribute of body, fits closely with Descartes's conception of the *quantitas materiae*. In his discussion of rarefaction and the vacuum in the *Principia*, Descartes writes:

> there cannot be more matter or corporeal substance in a vessel containing lead or gold or any other body, no matter how heavy and hard, than there is when it contains only air and is thought of as empty, for the quantity of the parts of matter does not depend on their heaviness or hardness, but solely on their extension, which is always equal in a given vessel. (*Principia Philosophiae*, VIII-1: 51)

The Cartesian view, then, is that each body of a given volume has the same extension and therefore the same *quantitas materiae*. To measure a body's quantity of matter, we measure a geometric property, namely its extension in three dimensions. And this geometric property appears to be

---

[24] Huygens provides a nice discussion of these issues in the preface to his *Discours* of 1690 – see *Œuvres complètes*, vol. XXI: 445–6 (this pagination is of vol. XXI, since the original pagination for the preface of the 1690 edition is not given marginally, as it is for the rest of the text).

[25] These views are discussed in part II of *Principia Philosophiae*, especially in sections 4–5, 10–12, and 18–19 (VIII-1: 42–3, 45–7, 50–2). Newton read and analyzed this part of the Cartesian text in depth in *De Gravitatione*.

invariant with respect to the "heaviness" of the body that bears the property. For Descartes, one cannot determine a body's quantity of matter by weighing it because weight is unrelated to the body's geometric properties.[26]

At least prima facie, one might think that Newton's *quantitas materiae* (or mass) differed from Descartes's in an obvious way: Newton takes the quantity of matter to involve not just extension, but the density of the extended body. Indeed, Newton explains what he means by the quantity of matter, or mass, in the first definition of the *Principia* by employing seemingly ordinary notions: "Quantity of matter is a measure of matter that arises from its density and volume jointly" (*Principia*, 403).[27] This gives us the picture of a volume as being "filled" with material. Newton discusses his definition further by supplying several ordinary examples: "If the density of air is doubled in a space that is also doubled, there is four times as much air, and there is six times as much if the space is tripled. The case is the same for snow and powders condensed by compression or any causes whatsoever" (*Principia*, 403). These comments suggest that Newton takes mass to be relatively straightforward.[28]

However, these first impressions may be misleading. Newton certainly does take mass to be an intrinsic feature of material bodies *per se*, one independent of a body's state of motion, but his conception of mass is more complex than his initial discussion suggests. To begin with, the notion of a volume being filled by some material seems vague,[29] not least because we

[26] Descartes also discusses the relation between weight and quantity of matter at *Principia Philosophiae*, VIIIA: 214–15. In proposition twenty-five of part IV of *Principia Philosophiae*, Descartes argues that bulk and gravity are not proportional to one another; Newton rejects this view in proposition 6 of book III of the *Principia*, where he argues that mass, bulk, and gravity are exactly proportional to one another (*Principia*, 806–9). I owe these last two citations to Nico Bertoloni Meli.

[27] As will become clear below, I focus primarily on what we would call inertial mass.

[28] In at least one pre-*Principia* manuscript, Newton presents the quantity of matter in a parallel fashion. In what may be Newton's first use of the term *massa* (mass), as opposed to the then common *moles* (bulk), the first definition in the *De Motu corporum* of 1684 or 1685 reads as follows: "*The quantity of matter* is that arising conjointly from its density and magnitude. A body twice as dense in double space is four fold. This quantity I designate by the name body or mass [*corporis vel massae*]" (Whiteside, *The Mathematical Papers of Isaac Newton*, vol. VI: 92; cf. also Herivel, *The Background to Newton's Principia*, 315, 317). The *Oxford English Dictionary* cites the first edition of the *Principia* (1687) as the first appearance of the word "mass" in the relevant technical sense; Newton never published *De Motu corporum* in any of its versions. In an earlier version of *De Motu*, which Whiteside dates to the autumn of 1684, the term *massa* and the attendant concept do not appear (cf. *The Mathematical Papers*, vol. VI: 30). And *massa* does not in fact appear in *De Gravitatione*, despite the translation in Hall and Hall's version (*Unpublished Scientific Papers*, 149), and in my own (*Philosophical Writings*, 37), for the original has *moles* and not *massa*. Thanks to George Smith for his assistance with these issues.

[29] This should not be conflated with the issue – which hails at least from Mach in the late nineteenth century – of the supposed circularity of Newton's definition of mass via density and volume. For a

lack a clear understanding of density. Is the density of some volume determined by the number of minute particles comprising it, by the weight of the substance occupying the volume, or in some other way? We can certainly follow Newton's calculation – in this case – but do we really understand the claim that the density of air in a space is "doubled"? Newton's few comments at this stage seem insufficient to guide our understanding.

Newton is presumably aware of this difficulty, so he renders his concept more precise, leaving aside the potentially confusing notion of density. For instance, in definition three he indicates that inertial mass is a measure of a body's resistance to acceleration (*Principia*, 404).[30] I take this to be central to Newton's conception of mass. He also provides a precise measure for the quantity in question; indeed, one of the *Principia*'s innovations is to provide measures for each of the quantities that are broached. In the discussion of the first definition, we are told that mass "can always be known from a body's weight, for – by making very accurate experiments with pendulums – I have found it to be proportional to the weight" (*Principia*, 404).[31] In this one sentence, we find two crucial elements of Newton's conception of mass. First, we can determine the unknown quantity, mass, by determining a known quantity, weight.[32] This establishes a crucial difference between the Cartesian and the Newtonian versions of the quantity of matter: Descartes denies that one can determine a body's quantity of matter by weighing it. Second, we learn that mass and weight are distinct quantities.[33]

discussion of Mach's critical interpretation, see Jammer, *Concepts of Mass in Contemporary Physics and Philosophy*, 11ff., and Cohen, "Newton's Concepts of Force and Mass," 59. With respect to Mach's question, one must realize that although Newton's usage of "density" appears to be ordinary, or non-technical, in fact it is a technical usage, one meaning specific gravity. Even in earlier texts Newton is clear that density in this sense is determined by placing an object in water – see Hall and Hall, *Unpublished Scientific Papers*, 168, and Cohen's editorial notes, *Principia*, 94–5. Thanks to George Smith for discussion of this point.

[30] In this definition, Newton defines what he calls *vis insita*, or the inherent force of matter. Since this is equivalent to the inertial mass of a body, as Newton says, it is not an impressed force as Newton understands that notion, and as he defines it in definition four (*Principia*, 405). Hence it is not a Newtonian force at all in the usual sense; thus in what follows, I ignore this terminology and write of *mass* instead. For a parallel discussion, see Cohen, "Newton and Descartes," 630; cf. also Blay, *Les "Princpia" de Newton*, 42–3. For different perspectives on this issue, see Jammer, *Concepts of Mass*, 80–7, and McGuire, "Natural Motion and its Causes."

[31] Bertoloni Meli (*Thinking with Objects*, 159) helpfully notes that the experiments can be seen as a response to Descartes's view of the relationship between *quantitas materiae* and weight in *Principia Philosophiae* (VIII-1: 214).

[32] To be precise, Newton contends that "quantity of matter" is strictly proportional to weight, but the pendulum experiments he cites here indicate only that inertial mass is proportional to weight (cf. *Principia*, 806–9). See also book II, proposition 24, corollary seven (*Principia*, 701). Many thanks to Michael Friedman for discussion of this point; he provides an illuminating analysis of these issues in *Kant's Construction of Nature*, 295–9.

[33] Whereas the mass of an object is an intrinsic feature of it, one independent of its spatial relations, weight is an extrinsic feature, one that depends on (what we would call) the gravitational fields

The pendulum experiments that Newton mentions in definition one are described in depth in book III, proposition 6; this section is evidently aimed at refuting Descartes. Proposition 6 itself reads: "All bodies gravitate toward each of the planets, and at any given distance from the center of any one planet the weight of any body whatever toward that planet is proportional to the quantity of matter which the body contains" (*Principia*, 806). Newton then describes his experiments as follows:

Others have long since observed that the falling of all heavy bodies toward the earth (at least on making an adjustment for the inequality of the retardation that arises from the very slight resistance of the air) takes place in equal times, and it is possible to discern that equality of the times, to a very high degree of accuracy, by using pendulums. I have tested this with gold, silver, lead, glass, sand, common salt, wood, water, and wheat. I got two wooden boxes, round and equal. I filled one of them with wood, and I suspended the same weight of gold (as exactly as I could) in the center of oscillation of the other. The boxes, hanging by equal eleven-foot cords, made pendulums exactly like each other with respect to their weight, shape, and air resistance. Then, when placed close to each other [and set into vibration], they kept swinging back and forth together with equal oscillations for a very long time. Accordingly, the amount of matter in the gold (by Book II, proposition 24, corollaries 1 and 6) was to the amount of matter in the wood as the action of the motive force upon all the gold to the action of the motive force upon all the [added] wood – that is, as the weight of one to the weight of the other. And it was so for the rest of the materials. (*Principia*, 806–7)

So it is an empirically established fact that (inertial) mass is proportional to weight.[34] Hence Newton indicates here that we can always measure the (inertial) mass of a body because of its proportionality to weight.[35]

---

generated by other objects. That is why Newton does not list weight among the "universal qualities" of bodies. Larry Sklar provides a helpful discussion of some of these issues in his "Physics, Metaphysics and Method in Newton's Dynamics." Intriguingly, Descartes distinguishes between weight and quantity of matter. He contends that the heaviness of bodies on earth is dependent on the fact that they are surrounded by other matter that impacts upon them, pushing them toward the earth's surface. If the same bodies were surrounded by other bodies that did not hinder their motion, their heaviness would disappear (*Principia Philosophiae*, VIII-1: 212–13), but this would do nothing to alter their quantity of matter. However, Descartes does not take quantity of matter to be proportional to weight.

[34] This conclusion amounts to the claim, in modern language, that inertial and gravitational mass have been determined empirically to be equal. This is evident already from the second law as it is now understood. Put briefly, and again in modern language, if ($f = m_i a$) and weight is measured by ($w = m_g g$), then since weight is a force, we can substitute and obtain: [($m_i = a$) = ($m_g g$)]. Since $a = g$, we find that $m_i = m_g$. Cf. also n. 35.

[35] In discussing proposition 6, Newton indicates (*Principia*, 806) that Galileo's results regarding free fall already suggest that inertial mass is proportional to weight – or in our terms, that inertial and gravitational mass are equal – for otherwise bodies with different (inertial) masses would exhibit different accelerations in free fall. He then notes that his pendulum experiments show this with even

Newton's rejection of the Cartesian view that extension is the essence of matter, and that the *quantitas materiae* must reflect this view, has another consequence. Newtonian mass and Cartesian *quantitas materiae* bear distinct relations to the other fundamental qualities of bodies articulated by Newton and Descartes, respectively. As we have seen Descartes contends that the quantity of matter of a body can be exhaustively understood in terms of its geometric properties; it depends on "extension alone" (*Principia Philosophiae*, VIIIA: 50–1, and 52–3).[36] So we might consider how mass relates to the other universal qualities Newton discusses in rule three (*Principia*, 795) – and it seems clear that a body's mass cannot be exhaustively characterized by its other qualities. To understand this claim we can rely on Newton's idea that (inertial) mass is a measure of a body's resistance to acceleration.[37] The idea is roughly this: a mobile, impenetrable, and extended body would resist any penetration of the space it occupied, and it would do so regardless of its state of motion, but it could certainly fail to resist acceleration. After all, if one such body collides with another, their extension and impenetrability prevent them from occupying the same three-dimensional spatial location at the same time, and their mobility indicates that they may change one another's motion. But this is perfectly consistent with their colliding and then accelerating arbitrarily. A mobile, impenetrable, and extended body need not follow Newton's laws.

This suggests a related point. Unlike Descartes's *quantitas materiae* – which is explicable in terms of extension alone – mass is not comprehensible in terms of Newton's other universal qualities. The fact, for instance, that a body is "mobile" is certainly bound up with – or perhaps amounts to – the fact that it can occupy distinct spatial locations at distinct times, and being "extended" and "impenetrable" are bound up with what it means for a body to occupy a space to the exclusion of other bodies.[38] But these notions do not involve the concept of acceleration, or even of a rate of change. An understanding of mass requires an understanding of Newton's laws of motion, especially of the concept of resisting acceleration. That is to say,

---

greater accuracy. Of course, Newton is not in a position to explain what we would call the equivalence of inertial and gravitational mass; that had to wait for Einstein's theory in the twentieth century. But note that Newton does explain the relationship between inertial mass and weight, indicating – in definition four – that inertial mass is always proportional to its bearer's weight (*Principia*, 404). Many thanks to Michael Friedman and George Smith for discussion of these issues.

[36] Cf. Garber's comment: "The bodies of physics are the objects of geometrical demonstration made real" (*Descartes's Metaphysical Physics*, 89).

[37] Newton emphasizes in his first and third definitions (*Principia*, 404) that mass is a measure of a body's "inertia," or of its resistance to acceleration. This retains currency today: Feynman characterizes mass as "fundamentally a measure of inertia" (*Lectures on Physics*, vol. I: 7–11).

[38] See the subtle discussion in Warren, "Kant's Dynamics."

even a complete understanding of the other Newtonian universal qualities of body is insufficient for achieving an understanding of Newtonian mass; the latter is attainable only through an understanding of the laws and definitions presented in Newton's physical theory (I discuss some implications of this view below). So unlike the Cartesian *quantitas materiae*, Newton regards mass as an irreducible feature of a material body, that is, a feature that does not reduce to its other properties, and that cannot be understood in terms of those properties. In that regard, we know not only that mass is an essential property of material bodies,[39] we know that it does not reduce to (what Newton takes to be) the other essential properties, namely, extension, hardness (inelasticity), and impenetrability. If mass did reduce to one or more of these other properties, then presumably it would be hasty to list it as one of the essential properties of bodies – at best, the list would be redundant.

Newton and Descartes also differ on the question of whether material bodies could consist of atomic constituents. Descartes, of course, attempts to refute atomism on *a priori* grounds in *Principia Philosophiae*,[40] but it would be a mistake, it seems to me, to understand Newton as simply endorsing the atomistic picture that Descartes rejects. Instead, Newton appears to be agnostic on the question, taking the issue of atomism to be an empirical one. It is well known that part of Newton's tactic in his discussion of rule three is to emphasize that the essential properties of material bodies – much like the primary qualities in Locke – characterize not only macroscopic bodies, but also their minute parts. Hence the microscopic constituents of material bodies, whatever those constituents may turn out to be, also bear mass. This obviously fits with Newton's view that gravity acts on all the "parts" of bodies and not merely on their surfaces. And for Newton, of course, mass is additive in the sense that the mass of the whole results from the mass of the parts, and increases as we increase the number of parts with a given mass. But it is crucial to see that this view is nevertheless agnostic on the question of atomism: in rule three, Newton does not contend that the microscopic constituents, or the parts, of bodies are themselves atomic entities.[41] In his discussion of rule three, Newton clarifies his view:

---

[39] See book II, proposition 40, Scholium (*Principia*, 761) and the more famous passage in the discussion of rule three (*Principia*, 796).

[40] See *Principia Philosophiae*, VIII-1: 51–2; for discussion, see Garber, *Descartes's Metaphysical Physics*, 120.

[41] One reason may be that the mature Newton takes the question of atomism to be an empirical one; hence the view that bodies have microscopic constituents that are in fact atomic is subject to empirical

Further, from phenomena we know that the divided, contiguous parts of bodies can be separated from one another, and from mathematics it is certain that the undivided parts can be distinguished into smaller parts by our reason. But it is uncertain whether those parts which have been distinguished in this way and not yet divided can actually be divided and separated from one another by the forces of nature. But if it were established by even a single experiment that in the breaking of a hard and solid body, any undivided particle underwent division, we should conclude by the force of this third rule not only that divided parts are separable but also that undivided parts can be divided indefinitely. (*Principia*, 796)

So "parts" refers agnostically to the insensible constituents of bodies that have thus far fallen within the range of our senses, and any other constituents that bodies may in fact consist of, regardless of whether they are atomic. His view would hold even if matter were infinitely divisible – in that event, at any arbitrary stage of division, the parts in question would bear mass, for mass is a continuous quantity and therefore has no smallest unit.

Since mass is a continuous quantity, the mass of a body can be measured to vary by any amount – in principle, there is no smallest amount of mass that can be measured, just as there is in principle no smallest acceleration, or smallest force, that can be measured. But this is perfectly consistent with Newton's agnosticism regarding atomism: even if there were atoms, Newton's use of rule three, as we have seen, does not entail that atomism is true, and there is no guarantee that we could measure the mass of an atom, so we might simply fail to discover that there was in fact a smallest mass.

Newton must still confront a mechanist objection to his overall view. As Newton notes explicitly in rule three, his theory does characterize the essence of matter, but through mass rather than the force of gravity. This raises an intriguing question: suppose that the mechanists had focused specifically on this view, dropping their contention that if gravity existed, it must be an occult quality, replacing it with the new claim that mass was an occult quality. After all, from the mechanist point of view, it may not be immediately clear that mass is a manifest quality of bodies – certainly the mass of a body is not evident upon a perceptual inspection of it in the way that the mechanist properties of extension, impenetrability, and mobility are. Perhaps Newton's method of avoiding the mechanist critique of gravity simply saddles him with the view that bodies have another occult quality.

---

disconfirmation. This represents an important conceptual shift from Newton's youthful view – one that owes a great deal to his reading of Henry More's discussion of atoms – where he seemed to think that one could prove the existence of atoms through *a priori* reasoning alone. See Janiak, "Space, Atoms and Mathematical Divisibility in Newton," which also cites the relevant sections of Newton, *Certain Philosophical Questions*.

Does mass – in either its inertial or gravitational guises – meet the technical definition of occult quality above? Is mass understood to cause effects in one or more other objects, so that it is exhaustively characterized by these effects?[42] Although we have seen how Newton treats mass and gravity in crucially different ways, his treatment of force helps once again to clarify his position, and especially his possible responses to criticisms of it.[43] That is, one can return here to the themes discussed in ch. 3, emphasizing that mass is a quantity that can be measured through a clear procedure that Newton outlines in definition one (and discusses later in book III, proposition 6). This seems to indicate that mass is not in fact characterized solely by its effects – we do not simply characterize mass by indicating (say) that it causes a body to resist acceleration, we measure the quantity (for instance, through its proportionality to weight). Hence we can immediately distinguish mass from dormative virtue, which cannot be measured, and which therefore must be characterized solely through its power to cause sleep. To unite the conclusions of this chapter and the previous one, then, we can say that gravity is not an occult quality because it is proportional to mass, and therefore not characterized solely through its effects, and that mass is not occult because it is a measurable quantity.

Although both force and mass are continuous quantities that have clear measures in Newton's hands, our knowledge of mass, from Newton's point of view, extends considerably beyond our knowledge of force in general, and beyond our knowledge of the force of gravity in particular. Whereas we do not know whether gravity is some kind of non-essential property of material bodies – perhaps it is a relational property that depends upon the existence of some kind of medium – or of the ether, we do know that mass is an essential property of matter. We might even say that we can provide both a mathematical and a complete physical treatment of mass: the former explicates the two measures of the physical quantity in question – through weight and resistance to acceleration – and the physical treatment indicates that precisely that quantity is in fact an essential property of material bodies. In this case we are able to answer at least two distinct ontological questions: first, we might say that mass is a (continuous) quantity; and second, we might add something that we cannot say in the case of gravity, namely that mass is also an intrinsic and essential property of material bodies. In that

---

[42] I leave aside the third criterion, since Newton certainly thinks that mass is distinct from, and irreducible to, the other qualities of its bearer, as we have seen.

[43] I owe my understanding of this problem and its solution to Jeff Barrett, Alan Nelson, and Kyle Stanford.

regard Newton's conception of mass can satisfy the ontological demands presented by his mechanist interlocutors in a way that his view of gravity cannot. Our knowledge of gravity is largely negative: we know that it is not mechanical, and that it is not essential to matter; but we do not know whether it is a non-essential property of some kind, or perhaps a property of the ether.

Yet the fact that our knowledge of the ontology of mass extends beyond our knowledge of the ontology of gravity does not mean that we have exhaustive knowledge of the former. Here, again, the contrast with Descartes is stark. Newton is quite clear in explaining that there are various aspects of mass that we do not understand, in part because we remain ignorant of many aspects of the nature of matter. For instance, what compresses bodies, or their constituents, so that they have mass as one of their properties? What binds bodies together? Do bodies have "pores" – spaces between their constituent particles that are completely empty of matter – or are they fully solid? Even if they do have pores, we can still ask whether the pores are empty – whether they are each a vacuum – or whether they are "full" of some type of subtle matter, such as an ether. We therefore remain ignorant of the ontology of body in many respects. This ignorance is reflected already in definition one: "For the present, I am not taking into account any medium, if there should be any, freely pervading the interstices between the parts of bodies" (*Principia*, 403–4). From Descartes's point of view, of course, this ignorance is intolerable: in *Principia Philosophiae* we prove that atoms cannot exist through purely *a priori* means, and we also know that bodies have no pores because empty space, at any physical level, is impossible (it is ruled out on independent grounds). In that sense, although Newton can answer a classic mechanist question about the ontology of mass, he is certainly not in a position to answer every Cartesian question about matter.

Newton's approach to heaviness also differs fundamentally from Descartes's. To underscore the novelty of Newton's approach to weight, or heaviness, consider this: he takes neither the Scholastic route, nor the well-trodden Cartesian or strictly mechanist route, but something that in a way borrows from each approach, although it differs crucially from each approach as well. With respect to the Scholastic approach, he of course denies that gravity or mass is an occult quality, and we have seen that he is on firm footing for each, and yet he does believe that we ought to attribute to bodies a property in virtue of which they are heavy. That is, he does add a property to the austere mechanist ontologies embraced by Descartes and Locke. If we endorse one of those austere ontologies, then we are forced to take one of two tacks: we can say, *à la* Descartes, that bodies have only the mechanist properties, and that therefore heaviness is not a property at all, but simply an effect of the impacts on the surfaces of bodies on earth from vortical or etherial particles, or from a swirling fluid; or we can say, *à la* Locke, that heaviness or gravity is superadded to matter by God, which means that although we cannot possibly fathom how matter could gravitate towards other matter, we cannot infer from this fact that it is beyond God's capability to endow matter with the property of gravity.

Newton rejects each approach by arguing that we must add a new property to the mechanist ontology if we are to have any hope of under- standing heaviness. We have to understand that the heaviness of a body is a result of the gravitational interaction between the body and earth (or any other attracting body); and to understand that interaction, we must realize that it is proportional to the mass of each body. On that basis, we can articulate the law of universal gravitation – only then will we be in a position to understand why all bodies fall as Galileo discovered, namely with the same acceleration, independent of the material constitution of the body.[44] In that sense, the heaviness of a given body, if it involves the acceleration of that body toward earth, is in fact independent of its mass. That is evident through the law of gravity itself, which requires us to understand that a mechanical account of gravity cannot work, since gravity is not mechanical. Therefore we know, solely from the law of gravity itself, that heaviness or any other gravitational interaction property cannot be mechanical. So the Cartesian account simply must be rejected. But we should not stray from this territory too far, because we do not want to attribute an occult quality to bodies, as the Scholastics may have done. Instead, we deny that we know that gravity is a property of them at all, and we assert that mass is a property of them, but of course it is not an occult quality. In that regard, Newton rather remarkably rejected both mechanism and Scholasticism, attributing a non-mechanical property to bodies without attributing a Scholastic occult quality to them.[45]

---

[44] Put in modern terms, when we substitute $f = ma$ into the law of gravity, we find the following: the acceleration that earth produces in any body is proportional only to the mass of earth, and independent of the mass of the falling body. See the section on non-mechanical matter below for details.

[45] Even when we limit our attention to the supposed paradigm case of a mechanical phenomenon – that involving the movements of the hands of a clock – Newton thinks that we will find that at least one crucial aspect of the phenomenon is not mechanically explained, and is perhaps even mechanically inexplicable. If the clock's movements are the result at least in part of a weight within it, then Newton contends that we lack a satisfactory mechanical explanation of the clock's movements on the grounds that we lack any kind of mechanical explanation of weight. Newton is willing to admit that one can

THE EPISTEMOLOGY OF MATERIAL OBJECTS

Newton's and Descartes's epistemologies of body differ starkly. The Cartesian conception of body outlined very briefly in the section above is known *a priori*, forming part of the metaphysical foundation for Descartes's physics. We could neither discover this conception of body to be false through any ordinary perceptual experience – we know all spaces to be full of body, despite any potential lack of perceptual acquaintance with bodies in given spaces in the world – nor could any development within physics involve such a discovery, for the metaphysics of body is presupposed by Descartes's physics. Hence the metaphysics is logically prior to the physics. This is equivalent to Descartes's conception of God, which is also presupposed by the physics outlined in *Principia Philosophiae*, and which allows Descartes to "derive" his first two laws of nature from God's immutability *a priori*. Hence we arrive at an *a priori*, metaphysical picture of body and of the laws that govern the principal interactions of body in time and space.

Now the most obvious way to articulate the distinction between this Cartesian conception of our knowledge of body and Newton's own would be to follow a long tradition of emphasizing the broadly empiricist aspects of Newton's view of matter and physical theory.[46] Some of Newton's rhetoric aligns well with this tradition – e.g. in his discussion of rule three, Newton writes:

For the qualities of bodies can be known only through experiments; and therefore qualities that square with experiments universally are to be regarded as universal qualities; and qualities that cannot be diminished cannot be taken away from bodies. Certainly idle fancies ought not to be fabricated recklessly against the evidence of experiments, nor should we depart from the analogy of nature, since nature is always simply and ever consonant with itself. (*Principia*, 795)

Newton goes on to say that each of the properties he takes to be universal, and indeed essential to matter – namely extension, mobility, hardness, impenetrability, and mass – is known to us "only through the senses."[47]

---

give at least a partial explanation of the clock's movements, if one brackets the question of what accounts for the weight of one internal piece of the clock; that is, one can explain various aspects of the clock's behavior without recourse to the weight within it – we can say, for instance, that the minute hand moves because it is attached internally to a lever, and the lever moves because it is attached to a rotating gear, etc. That having been said, however, the mechanical philosophy leaves something crucial unexplained. See Newton's unpublished letter to the editor of the *Memoirs of Literature* of May 1712, mentioned above – *Philosophical Writings*, 116–17.

[46] This aspect of Newton's work is especially emphasized in Stein, "Newton's Metaphysics."

[47] For a sophisticated discussion, see Stein, "On Locke, 'the Great Huygenius, and the incomparable Mr. Newton,'" esp. 29–30; Stein's view in this paper bears some similarities to my own.

Certainly this view conflicts with the Cartesian view that I mentioned above. But the complexity of his concept of mass indicates, I think, that Newton's epistemology of material bodies is in fact more complex than it appears.[48] As in ch. 2, when I emphasized Newton's focus on refinement and revision as the key epistemic issues, here too I emphasize what I take to be Newton's own epistemic focus.

Leaving aside the other properties of matter, in what sense can we know that bodies have mass "only through experiments," or perhaps, only through perceptual experience? An obvious route to answering this question is to construe our perception of a body's mass on the model of our perception of the mechanist qualities discussed, say, by Locke. That is, we might suggest that we can perceive a body's mass through ordinary experience just in the way that we can perceive a body to be solid, extended, or mobile. And for Newton, as we have seen, we can understand a body's (inertial) mass as involving its resistance to acceleration, and we can measure it through its proportionality to weight. So, for instance, we might say that in feeling a body's resistance to acceleration, I feel its mass. But this raises questions that are absent in the case of the mechanist properties.

Here is one reason to be skeptical that we can perceive mass through what we might call ordinary perceptual experience. It seems that in ordinary parlance we do not distinguish between the following two things: resisting motion, and resisting acceleration. That is, we certainly have an intuitive sense that it takes effort to move a resting body – say a rock resting on the ground – or to stop a moving body – say a baseball flying towards us – but I doubt that we ordinarily think in terms of a body's acceleration, rather than its motion or its speed. It is questionable that we distinguish between speed and velocity, and since acceleration is the rate of change of velocity, and velocity is vectorial rather than scalar, it seems that in ordinary affairs we probably do not think clearly in terms of resistance to acceleration. As a result it is not clear that our ordinary experience of the effort it takes (say) to push a resting book across a table is an experience of a resistance to acceleration, rather than simply to motion. And when we stop (say) a grocery cart rolling down a hill, it seems we may experience the resistance simply as a resistance to being stopped, rather than as a resistance to acceleration. As these two examples already indicate, it would take considerable effort to show that we are in fact experiencing mass.

The same can be said of Newton's empirical method of measuring mass through its proportionality to weight. When we perceive ordinary material

---

[48] Once again, I focus especially on what we would call inertial mass, since this is Newton's focus.

objects such as rocks or books, their weight certainly seems to be an intrinsic feature, as it is presented in our experience; that is, it certainly does not appear to differ from location to location, regardless of whether one is on a mountain or in a valley, and regardless of the other bodies that lie within its vicinity.[49] It therefore takes some basic theoretical work to realize that the weight of a body is not an intrinsic feature of it, but rather a relational feature, one that increases and diminishes with changes in the bearer's relations. However, even with knowledge of these facts about weight, our perceptual experience would appear nonetheless to represent weight as an intrinsic property of a body. But of course, mass and weight are distinct for Newton: the former is an intrinsic feature of a body, independent of its spatiotemporal location, and the latter is a relational feature of a body, dependent on its relation to other bodies. In order to perceive mass we must perceive a property of a body that is proportional to, but distinct from, its weight. And that is precisely what is unavailable to us in ordinary perceptual experience.

It is unclear, then, that we ordinarily perceive the mass of an object, either through perceiving its resistance to acceleration, or through perceiving its weight. Certainly, the case of mass appears to be more complex than the case of the mechanist properties Newton also takes to characterize all material bodies. If you like, in the case of mass we have a series of possible confusions – between mass and weight, velocity and speed, and so on – that are simply absent in the case of solidity, extension, and mobility. In our ordinary experience of an object there is no danger that we might confuse our perception of its extension in three dimensions (say) with some other property it bears, such as its ability to occupy distinct spatial locations at different times, or its resistance to being broken apart.

It is precisely these complexities in the case of mass, it seems to me, that underscore the problems with the traditional interpretation of Newton's epistemology as straightforwardly empiricist in character. It is not clear that all the properties of material objects are discoverable through the senses, if by that we mean, through ordinary perceptual experience alone.[50] That is what some of Newton's interlocutors would have meant. Rather, it seems that our perception of a body's mass is also at least partially dependent on our knowledge of the concepts in the *Principia* – hence our perception must

---

[49] See Holton, *Introduction*, 66 for some relevant empirical data.

[50] That is what some defenders of the mechanical philosophy emphasized – see Wilson, *Ideas and Mechanism*, 477–8; I discuss Wilson's view briefly in the next section below. This perspective on the qualities of body might reasonably be found in some of Locke's reflections. See, for instance, his discussion of primary qualities at *Essay*, 2.8.9.

be aided by what we might call background knowledge, a kind of epistemic component that is missing in the case of the mechanist properties.[51] One can be ignorant of Euclidean geometry and still perceive the extendedness of an ordinary body, but one cannot be entirely ignorant of classical mechanics and still perceive the mass of that body.

The fact that we require background knowledge to perceive mass does not mean that our ordinary perceptual experience is simply useless; rather, it must be rendered more precise through the introduction of distinctions that we do not ordinarily make. Hence we have an intuitive understanding that it takes effort to lift an object – bodies are heavy – and that it takes effort to move an object. But those notions must be rendered more precise: we have to be clear that mass and weight are distinct and that speed and velocity are distinct. Once we have these concepts available to us, along with the precision that they allow, perhaps we can in fact perceive a body's resistance to acceleration, and not simply its resistance to motion or to rest.

But this interpretation remains vague. How much is contributed through our perception of a body, and how much by Newton's physical theory? We might say that mass is a semi-technical notion, which is to say, a notion explicable only in terms of concepts available in physical theory, but one that renders ordinary concepts and articulations of our perceptual experience more precise.[52] Mass is semi-technical because it cannot be characterized, or understood, independently of Newton's physical theory.[53] For Newton, to define mass is to provide a measure for it, as we have seen, and we cannot explicate mass's two measures without relying on the

---

[51] See Hanson's famous discussions of observation and "seeing as" in ch. 1 of *Patterns of Discovery* and in *Perception and Discovery*, 91–110.

[52] The question of whether the concept of mass is technical as opposed to non-technical should not be conflated with Carnap's question of whether it is "theoretical" as opposed to "operational." Newton provides what Carnap would probably call an operational understanding of mass; but from my point of view, it is crucial that the concept is bound up with other concepts in Newton's theory. Cf. Carnap, *Introduction to the Philosophy of Science*, 103ff. For a discussion of Carnap's view in the context of twentieth-century attempts to define mass, see Jammer, *Concepts of Mass in Contemporary Physics and Philosophy*, 20ff.

[53] With respect to the relation between what became the first law and the concept of mass in the *Principia*, Newton had something akin to the first law – or at any rate, he held a version of the principle of inertia – long before he had developed the *Principia*'s concept of mass. For instance, in his *Wastebook* (from 1665), Newton already has something close to the first law, but he does not yet have the concept of mass. So the link in Newton's mature thought between the first law and the property of material bodies whereby they resist acceleration took some effort for Newton to formulate – it did not come along simply with the formulation of the principle of inertia (cf. Westfall, *Force in Newton's Physics*, 343–4). It is this link, in my view, that helps to illuminate Newton's mature conception of matter: the properties of material objects cannot be articulated, or understood, independently of the laws in physical theory. Newton's early thought on these matters underscores the fact that this mature view requires a step beyond the articulation of those laws themselves.

concepts of weight and of acceleration in Newton's theory. For that reason, we cannot perceive the mass of an object without background knowledge.

However, mass is not a thoroughly technical notion. A thoroughly technical notion is explicable only in terms of a given theory's concepts, but it is also completely divorced from ordinary concepts and ordinary perceptual experience – hence it does not render them more precise. To see what I have in mind, consider the concept electron: to know what an electron is, you must understand what an atom is, what charge is, what rest mass is, and so on. But in addition, the ordinary concepts and experiences we employ to explicate electrons are mere heuristic devices. We think of electrons as tiny balls spinning around other tiny balls at tremendous speeds, but of course that is not what they are, and even our most general notions, such as spatial position, are useless for understanding them.[54] To have the concept electron is not to have a very precise idea of a tiny rotating ball.

We are now in a position to return to Newton's discussion of the properties of matter in rule three and in proposition 6 of book III in the *Principia*, indicating how we might reinterpret his seemingly naïve empiricist rhetoric. In contending that the properties of matter are known "only through experiments," Newton is not embracing a naïve-empiricist view, but rather a sophisticated twofold conception of the epistemology of matter: first, he is obviously denying the Cartesian view that we can determine the universal properties of matter purely *a priori*, or through reason alone; and, second, he is contending that experiments guided by the concepts in physical theory are required for us to determine the properties of matter. Just as we cannot simply deduce *a priori* that bodies are massive, we cannot simply rely on ordinary perceptual experience to discover this property. Instead, we must conduct experiments – such as the pendulum experiments Newton himself cites when discussing the proportionality between mass and weight – in order to determine the properties. As Newton writes in definition one, which defines mass: "It [mass] can always be known from a body's weight, for – by making very accurate experiments with pendulums – I have found it to be proportional to the weight" (*Principia*, 404). As we have seen above, these experiments are then described in detail in proposition 6 to book III, which indicates that the weight of a body is proportional to its mass

---

[54] Consider a standard source for introducing students to the concept of an electron: Feynman says that we ought to conceive of an electron in terms of a "probability cloud" (*Lectures on Physics*, vol. I: 6–8ff.).

("quantity of matter").[55] Unless we have some of the basic concepts of Newton's theory, we presumably cannot make sense of these experiments with the bob of a pendulum, in which he places numerous types of materials in the bob, recording the behavior of the pendulum with each type (*Principia*, 806–7). If we cannot distinguish mass and weight, for instance, we cannot run this experiment, nor can we fathom it. But with the conceptual background knowledge in place, we can experimentally determine that mass is proportional to weight. Only then do we understand mass. Therefore, we cannot determine through unaided perceptual experience alone what the properties of material bodies are, any more than we can do so through reason unaided by perceptual experience. Rather, we can reach this determination only through perceptual experience guided by our empirically based physical theory. That is, I think, a remarkable view in the early eighteenth century.

This interpretation, of course, raises a much broader question. Suppose that in rejecting the Cartesian view of our knowledge of matter, Newton does not embrace a naïve-empiricist conception of that knowledge, as I have suggested. To what extent does the resulting epistemic view set Newton on a collision course with the prevailing mechanical philosophy of his day? In denying that gravity is essential to matter and putting mass in its place, one might think that Newton had sought reconciliation with the mechanists, simply altering their view by adding one more property to their basic ontology. But as we will see, the complexity of mass indicates that Newton's concept of matter involved a fundamental rejection of the mechanical philosophy.

A NEW CONCEPT: NON-MECHANICAL MATTER

Since the mechanical philosophy involved various ideas and commitments throughout the seventeenth and early eighteenth centuries, to gauge whether Newton adopts a mechanist picture of matter we have to distinguish various aspects of that picture, or perhaps more accurately, various conceptions that were defended by one or another adherent of the mechanical philosophy in that era. We can distinguish, very roughly, between four

---

[55] Pendulum experiments are also described in proposition 24 of book II, where Newton notes in his fifth corollary that "universally, the quantity of matter in a bob of a simple pendulum is as the weight and the square of the time directly and the length of the pendulum inversely" (*Principia*, 701). He then adds in corollary seven: "by making experiments of the greatest possible accuracy, I have always found that the quantity of matter in individual bodies is proportional to the weight."

related but distinct views that might be characterized as "mechanist." As we will see, Newton rejects essential components of each mechanist view.

First, we might construe the mechanical philosophy as an ontological conception with an overarching view of causation: bodies are characterized solely by size, shape, and motion (and maybe solidity) and they interact solely through impact.[56] Second, we might emphasize Edwin McCann's influential interpretation of Locke, construing mechanism as an ontological conception coupled with an explanatory claim. He finds this view in a passage from Locke's *Elements of Natural Philosophy*, which reads as follows: "By the figure, bulk, texture, and motion of these small and insensible corpuscles, all the phenomena of bodies may be explained."[57] Third, we might follow Michael Ayers's suggestive, if controversial, conception of mechanism: it holds that the laws of physics can in principle be explained by being deduced from the attributes possessed essentially by all bodies.[58] If we emphasize the ontology mentioned in the first construal above, then on Ayers's view the mechanist would hold that we can explain the laws of physics solely through size, shape, and motion (and perhaps solidity). We might also bracket this ontology, allowing the Ayersian mechanist to hold the more general view that the laws of physics can be explained by the essential properties of material bodies, whatever those properties might be. Fourth and finally, following a suggestion of Margaret Wilson's,[59] we might take a mechanist to think that the primary qualities of body – viz. size, shape, motion, and perhaps solidity – are discovered through ordinary perceptual experience, and that any laws of nature must be at least consistent with the fact that bodies and their constituents bear these and only these qualities. Remarkably, Newton's concept of matter involves a rejection of the mechanical philosophy in all four of these guises.

Consider the first construal of mechanism. As we have seen already in ch. 3, Newton rejects the mechanist view that all causation involves surface action among material objects characterized by size, shape, and motion (and perhaps solidity). His view that the force of gravity is a non-mechanical cause – because it is proportional to mass – can be found, *mutatis mutandis*, in his conception of matter itself. For mass just is the only intrinsic feature

---

[56] Cf. Westfall's formulation in *Force in Newton's Physics*, 377, and Margaret Wilson's discussion in *Ideas and Mechanism*, xiii–xiv.

[57] See McCann, "Lockean Mechanism," 209.   [58] Ayers, "Mechanism, Superaddition," 210.

[59] Wilson mentions the first clause; I have added the second. She writes: "Seventeenth-century mechanists largely conceived of the qualities of the insensible entities to which their explanations appealed in terms of qualities 'given' in ordinary experience of big, perceivable bodies: especially size, shape, and motion" (*Ideas and Mechanism*, 477).

of any body in virtue of which it experiences gravitational interactions with other bodies, so it is in virtue of possessing mass, in turn, that bodies undergo non-mechanical causal interactions with one another.[60] In that regard, Newton's non-mechanical conception of causation – according to which there can be local action that does not involve impact – coincides with his non-mechanical ontology, thereby mirroring the mechanist picture he eschews. Thus in this regard, Newton takes matter to be non-mechanical. If you like, the fact that bodies are massive means that bodies interact with one another non-mechanically.

A caveat is in order here: mass is essential to matter, and in virtue of mass matter is non-mechanical, but that does not mean that matter is essentially non-mechanical. For gravity is not essential to matter. Gravitational interactions between material bodies may in fact depend on the ether between them, so massive bodies in the absence of the ether might fail to gravitate toward one another. As we have seen, the *Principia* certainly cannot rule out such a possibility. But to conclude that gravity is essential to matter, one must be in a position to rule it out. So a more precise formulation of Newton's view would be as follows: matter in our world, with its "physical cause" of gravity in place (whether it is the ether or another medium), is non-mechanical; but matter is not essentially non-mechanical, even though matter in our world is non-mechanical because of one of its essential properties.

This raises a further question: if mass were the only intrinsic feature of a body in virtue of which it experienced gravitational interactions, how could the bodies in the ether-less world bear mass and yet lack gravitational interactions? As in other cases, the equivalence of what we would call inertial and gravitational mass is useful here. We can conceive of the bodies in the ether-less world as bearing mass in the sense that they follow the laws of motion.[61] Hence if they were moving rectilinearly in (absolute) space, they would continue to do so indefinitely. That is, they would resist acceleration, even if there was no gravity in their world because its physical cause had been removed (say by God).[62] Thus we would envision a world in which the bodies resisted acceleration, but in which they were not heavy.

Newton's view that mass characterizes the essence of matter stands in contrast with the influential and controversial understanding of gravity articulated by Locke. In his correspondence with Stillingfleet, Locke contends

---

[60] Here of course I refer to what we would call gravitational mass. I owe this formulation to discussions with Alan Nelson.

[61] I owe this point to Michael Friedman.

[62] Of course, this imagined world may require us to think of its inhabitants as occupying absolute space, and that raises questions of its own, but I leave those aside until ch. 5.

that God "superadded" gravity to matter, just as God could, or may very well have, superadded thought to matter. Locke is concerned with the fact that we cannot conceive of how gravity could be essential to matter, or could flow from the essential properties of matter, where he takes matter to be extended solid substance.[63] In his second reply to Stillingfleet, he writes:

> But I am since convinced by the judicious Mr. Newton's incomparable book, that it is too bold a presumption to limit God's power, in this point, by my narrow conceptions. The gravitation of matter towards matter by ways inconceivable to me, is not only a demonstration that God can, if he pleases, put into bodies, powers and ways of operations, above what can be derived from our idea of body, or can be explained by what we know of matter, but also an unquestionable and every where visible instance, that he has done so. And therefore in the next edition of my book, I shall take care to have that passage rectified. (Locke, *Works*, vol. III: 467)

From Newton's point of view, of course, if we leave mass out of our conception of matter's essence, and consider bodies to be essentially mechanical – that is, to bear essentially size, shape, and solidity alone – then Locke is correct in thinking that it is impossible to conceive of how a material object could interact gravitationally with another material object on the basis of its essential properties. For their interaction would be proportional to two quantities – mass and distance – but they would be characterized by only one of them. If we take bodies to be massive, however, then we may have a reason for avoiding the route of superaddition: if bodies are essentially massive, and if they can be spatially separated from one another, then we can understand how their gravitational interactions can be measured by measuring one of their essential properties, and another property that fits with their other essential properties. The principal remaining question then concerns the medium – the "physical cause" – that enables gravitational interactions without allowing action at a distance into natural philosophy. But at least one of the reasons for allowing the possibility that God superadded gravity to matter would be removed.[64]

---

[63] I have been influenced by conversations with Lisa Downing concerning the best interpretation of Locke's doctrine of superaddition.

[64] There is another contrast between Locke's and Newton's views here that is illuminated by the discussion of superaddition. In the passage from his letter to Stillingfleet quoted above, Locke seems to endorse the following view. Our conception of matter is fixed independently of physical theory and its discovery of gravitational interactions between objects. For Locke, Newton's theory tells us that bodies have a power to interact with other bodies that is inexplicable by our concept of matter; the theory does not indicate what matter is, or what a material body is. For Newton, however, it seems that our concept of matter is partially dependent on physical theory, so that if our theory tells us about a fundamental or universal power of bodies, that itself could alter our concept of matter, becoming part of the concept.

In one regard Newton and Locke agree on a significant entailment of the theory of gravity in the *Principia*: matter is non-mechanical, because it gravitates. For Locke, of course, this is inexplicable because matter is essentially just extended solid substance; thus matter is not mechanical because God superadds "gravity" to it. For Newton, in contrast, it is precisely because his theory of gravity requires us to think of matter as massive, and because he takes mass to be essential to matter, that matter is in fact non-mechanical. And this conception of matter then allows us, in turn, to begin the proper explication of gravitational interactions.

Newton's response to the second construal of mechanism will follow similar lines. Recall that on this construal, the mechanist holds the following Lockean view, which McCann finds in the *Elements of Natural Philosophy*: "By the figure, bulk, texture, and motion of these small and insensible corpuscles, all the phenomena of bodies may be explained." As should be evident from the above discussion, the key question here is whether Locke has anything akin to Newtonian mass in mind when he lists bulk among the properties of corpuscles. In a provocative recent essay, Woolhouse argues that Locke does in fact have Newtonian mass in mind.[65] As Woolhouse indicates, in the *Elements of Natural Philosophy*, Locke defines the "quantity of motion" as follows: "The quantity of motion is measured by the swiftness of the motion, and the quantity of the matter moved, taken together" (Locke, *Works*, vol. II: 416). However, it remains unclear that Locke has Newtonian mass and momentum in mind in this definition, since he does not clarify whether his quantity of motion is a scalar quantity, as in Descartes, or a vector, as in Newton; his "swiftness" of motion seems ambiguous between the two. This reflects the fact, in turn, that Locke is not quite clear whether the quantity of matter involves a resistance to motion, or to acceleration. Locke also apparently fails to recognize that mass has two measures.[66] Hence it is doubtful that Lockean bulk amounts to Newtonian mass, and it is therefore doubtful that Newton can accept the conception of mechanism emphasized by McCann. Otherwise put, Newton's view demands that bulk refer to mass, and it is unclear that Locke would endorse that demand.

Regarding Ayers's (the third) conception of the mechanical philosophy: does Newton think that the laws of physics can in principle be explained by being deduced from the essential attributes of all bodies? The primary mechanist route to providing this explanation, it seems, lies in becoming

---

[65] See Woolhouse, "Locke and the Nature of Matter," 151–3.
[66] Thanks to Lisa Downing and Tad Schmaltz for discussion of this point.

acquainted with the essential attributes of material bodies through ordinary perceptual experience – by which I mean, minimally, through perception independent of any knowledge of the laws of nature – and then harnessing our understanding of these attributes to explain the laws. Hence our ordinary perception acquaints us with the size, shape, motion, and perhaps solidity of bodies, and these attributes enable us to explain the laws of nature by deducing the laws, at least in principle, from these attributes. Perhaps the idea is that, at least in principle, we can explain why the laws of nature govern the interactions of material bodies by pointing to the essential attributes of those bodies. For instance, if the laws are laws of impact, or make reference to impact, we might emphasize that they govern bodily interactions because bodies are essentially extended and solid, and therefore can interact with other bodies through impact.[67]

Newton clearly rejects the mechanist view that Ayers identifies. From his perspective, we cannot determine the primary qualities, or the essential properties, of bodies independently of the laws of nature, because that would presumably leave mass out of our conception of matter. For (inertial) mass is defined clearly through the three laws of motion, and can be understood and then recognized to be a property of material bodies only through knowledge of the laws.[68] And I have suggested that we cannot perceive bodies to have mass unless Newton's laws form part of our background knowledge. In that regard, it seems that we cannot deduce the laws of nature from the essential attributes of body unless we already know the laws, in order to be able to recognize those essential attributes.[69]

[67] Of course, whenever an *in principle* clause is in effect, one wonders whether the adherents of the view modified by the clause actually thought that they, or perhaps even their followers, had any real hope of obtaining the knowledge in question. But we can bracket this issue here, for as we will see, Newton's disagreement with the mechanists is sufficiently general to render the *in principle* clause irrelevant.

[68] Newton characterizes the three laws of motion in this way in query 31: *Opticks*, 397.

[69] When presenting a possible way in which God could create matter by endowing parts of space with various properties in *De Gravitatione*, Newton does not list mass as one of the primary properties of body (*De Gravitatione*, 28–9). In that regard the view I attribute to him in this chapter is not reflected in this pre-*Principia* manuscript, but must await Newton's development of the concept of mass, which can probably be dated to the mid-1680s (see the discussion of *De Motu Corporum* above). But in *De Gravitatione* Newton already articulated part of the kernel of his later view, for he clearly stated that we would take something to be a body only if its properties were nomothetic – hence to be "mobile" is to follow what we take to be the laws of motion. That is, if a body were to suddenly start moving on its own, to move arbitrarily, etc., we would not consider it to be an ordinary body. The same goes for extension and impenetrability; for instance, if the body were hard to my touch but not to yours, that would not count as nomothetic impenetrability. What he does not add here, however, is the further claim that to understand what mobility or extension is, we must have the laws of nature. And in the *Principia*, it seems, the implication is that we would not understand mass independently of the laws of motion. That is a crucial further step.

We might also note that Newton lacks Descartes's mode of explaining his first two laws of nature, namely indicating how they follow from God's immutability.[70] From Descartes's perspective, these laws of nature are fully intelligible – we understand precisely why the laws of nature have the content that they have, for they follow from an essential attribute of the divine being that decreed the laws, and indeed created all the material bodies that obey the laws. We can use the laws to explain any event in the world, and we can use our knowledge of God to explain the laws. Thus there is nothing left unexplained about our world and the facts that characterize it, including its laws of nature – hence they are fully intelligible. This is distinct from the mechanist route to explaining the laws of nature that Ayers identifies, but it provides another way to achieve a full explanation of the laws.

Of course, Newton concurs that God created both the material world and all the laws that govern it, including any laws governing forces that we have yet to understand – such as electricity and magnetism, from Newton's perspective – or yet to discover. And Newton argues explicitly in his correspondence with Bentley in 1693 that we cannot explain the structure of our solar system – e.g. the various positions of the planetary bodies and their satellites – through reference to the laws of nature and the properties of matter alone; rather, we must appeal to God's wisdom and intelligence to achieve that explanation. But Newton never endorses Descartes's view that we can deduce the laws of nature from God, or from God's attributes. Rather, he takes our knowledge of the laws of nature to be empirical; we cannot know them from "reason alone" because we cannot deduce them from our knowledge of God, or from any *a priori* knowledge we might have. Thus from Newton's point of view, we cannot answer the question of why the laws of nature have the content that they have – we can only discover them through empirical research, and our empirical findings presumably can never make the laws fully intelligible in the way that Descartes seeks. In general, then, it appears that from Newton's point of view we simply lack the kinds of explanation of the laws of nature found in the Ayersian mechanists in general and in Descartes in particular.

As we have seen above, Newton takes the laws of nature to be empirical, and in several senses. One sense is that our knowledge of them is subject to refinement, revision, and even outright rejection, which is presumably not the case with the Cartesian laws. But there is another significant sense: Newton thinks that the laws are not conceptual truths. But in what sense are

---

[70] See the helpful discussion in Hatfield, *Descartes and the "Meditations,"* 297–8.

the laws of nature not conceptual truths? After all, the class conceptual truth might simply be understood as empty for Newton, in which case it would be trivial to say that natural laws were not members of that class. But from my point of view, it is not clear that the class is in fact empty for Newton. Consider the difference between Newton's conception of the laws of nature and his view that action at a distance is "inconceivable." It seems that for Newton, there is at least one basic conceptual constraint on our under-standing of material bodies and their interactions, but this constraint is not imposed by the laws of nature. As far as Newton is concerned, we can imagine a world in which the laws of motion do not hold, and therefore one in which the law of gravity also fails to hold. Imagine a world consisting solely of two small spheres, one not governed by Newton's laws. To envision this world is to conceive of these two bodies as massless, since to be massive just is to accelerate proportionally to impressed force, and therefore to be massive is to follow Newton's laws (and to follow Newton's laws is to be massive). So the idea is that we can imagine two extended, impenetrable, mobile bodies that are lonely in their world. Thus the laws of the *Principia* are not conceptual truths in the sense that we can envision a world consisting of bodies in which the laws do not hold.[71]

But can we also imagine that these two lonely spheres act on one another at a distance? Can we imagine that they move one another without impact-ing upon one another, and independently of any medium of any kind between them (*ex hypothesi*, they are lonely, so no medium exists)? For Newton, we cannot: it is inconceivable that material bodies can act on one another at a distance. In that regard, although Newton's laws, and his law of gravity, are perfectly consistent with distant action, Newton takes the latter to be impossible in the sense that it is ruled out independently of the laws of nature. You can, of course, imagine that two lonely bodies accelerate arbitrarily in their world, but you cannot imagine that their accelerations are due to their action on one another when they are not in contact. You could imagine that one moved, or jiggled, and that the other – spatially separated from the first – then moved or jiggled, but that would not be a

---

[71] As for the question of whether Newton would allow that we could envision bodies that were not massive – and therefore not material, given his view that mass is essential to matter – consider the fact that Newton himself does precisely that in his *De Gravitatione*, in which he takes bodies to be created through the divine imposition of nomothetic impenetrability and mobility on areas of extension (28–9). As far as I can tell there is no evidence that Newton altered his understanding of this creation story later in his life, and he even may have cited precisely this story years later when asked by Pierre Coste to explain what Locke may have had in mind by referring obliquely to a possible way in which God could be understood to create matter. See Metzger, *Attraction universelle*, part I: 32, and Stein, "Newton's Metaphysics," 272–3, for details.

way of imagining action at a distance.[72] That would not be a way of imagining distant causation, but rather only a way of imagining correlated motions.

Newton's response to the fourth construal of the mechanical philosophy mirrors his response to the third. He denies that we can determine the essential properties of material bodies independently of physical theory. We can fully determine the essential qualities of material bodies only via physical theory itself – without the latter, our conception of matter would be partial, leaving out mass. But the view has a further nuance. In determining the qualities of bodies, we do not thereby invent qualities that are (what I have called) thoroughly technical. Rather, as we have seen, physical theory defines certain qualities, such as mass, by beginning with ordinary concepts and ordinary experience and rendering them more precise. This is a significant nuance, in turn, because it indicates the subtlety of Newton's view. If he thought that we could not fully determine the essential properties of material bodies independently of physical theory because the latter postulated thoroughly technical properties or material entities, his view would be somewhat unsurprising. He would be calling in that case for a replacement of what Sellars calls the manifest image of objects with the scientific image.[73] He might be saying, for instance, that the actual properties of bodies and their constituents are nothing like the properties that are evident in our ordinary perceptual experience of bodies. And in that case, of course, we would have fundamentally rejected the mechanical philosophy, but we would have done so at the cost – if it is a cost – of severing the link between our manifest and our scientific conceptions of bodies, substituting the latter for the former.

Newton never severs that link. He occupies a crucial middle position between two extremes.[74] He does not reject the mechanical philosophy – and its view that the laws of nature should ultimately be explicable in terms of the essential properties of matter – by replacing it with the stark view that all the properties of material bodies are thoroughly technical in character, and that therefore no properties can be determined independently of physical theory. That view would tell us that the properties of bodies we ordinarily perceive are not their real properties – the latter would be technical properties available to us only through physical theory itself (a situation that may characterize early twentieth-century physics). Instead,

---

[72] Thanks to Michael Della Rocca for an imaginative discussion of this issue.
[73] See Sellars, "Philosophy and the Scientific Image of Man."
[74] Cf. especially the discussion in Wilson, *Ideas and Mechanism*, 455–94.

Newton occupies a middle position between these extremes: he rejects the mechanical philosophy without inverting it, contending that some of matter's properties are semi-technical, which is to say, explicable in the terms available to us through physical theory. But those properties are semi-technical in the sense that they are also available to us through ordinary perceptual experience; the latter must simply be rendered more precise. From Newton's point of view, then, the manifest image of bodies is both incomplete and imprecise, but the scientific image of bodies begins with the manifest image, supplementing it and rendering it more precise.[75]

But Newton does not merely endorse a novel, non-mechanical conception of matter. That conception, in turn, can serve as the basis for an independent critique of the mechanist conception of matter. Two criticisms of the mechanical philosophy (broadly construed) are salient here: first, it is unclear that the mechanists can unify celestial and terrestrial phenomena in the way that Newton can if they fail to attribute mass to material bodies. For it is through that attribution that we can understand both material bodies on earth, and the planets and satellites, to be governed by the law of gravitation, and that, in turn, allows us to achieve the unification in question.[76] The mechanists may have been committed to treating the "sub-lunar" and "super-lunary" realms through the same physical principles, attributing to bodies throughout the universe the same basic mechanist properties, and therefore the same responses to the physical laws, but they appear unable to follow Newton's unification. In that regard Newton's concept of mass plays a crucial role in achieving a physical theory that explains phenomena throughout the world, a goal that mechanists from Descartes onward strongly endorsed, but never achieved.

Leaving aside the unification of celestial and terrestrial phenomena under one theory, we can focus solely on the case of bodies near the earth's surface to articulate the other Newtonian criticism of the mechanists. From Newton's point of view, the mechanist failure to attribute mass to bodies leads to the mechanist failure to understand and explicate a basic phenomenon first discovered by Galileo, and later recognized as greatly significant by many of the major natural philosophers of the later seventeenth century,

---

[75] I discuss an important parallel to this aspect of Newton's thought, very broadly construed, in ch. 6 below. Michael Friedman has articulated and defended a similar view in the context of his work on Kant. For instance, in a recent paper, he writes: "The objects of universal gravitation . . . are perfectly accessible to us . . . and the theory of universal gravitation thus serves to provide what is essentially a more exact and rigorous description of objects that are already present in everyday life" ("Kant on Science and Experience," 265). His view has influenced my own.

[76] My discussion of this unification parallels – and has been influenced by – Friedman's discussion in *Kant's Construction of Nature*, especially 292–5.

namely the fact that all bodies on earth fall with the same acceleration. If the mechanists were right that all material bodies are characterized solely by size, shape, motion, and solidity, why should all bodies in a vacuum fall with the same acceleration?[77] One would presume that on the mechanist picture, bodies of differing sizes or shapes would fall with different accelerations, for if free fall were due to the impact of material particles, or some kind of matter, on the surface of a falling body, would not a larger surface area produce a greater acceleration? This could be especially pressing if two bodies of two considerably different sizes – with considerably different surface areas – were dropped from the same height in the same vacuum chamber.[78] What property of bodies could a mechanist appeal to in order to avoid the conclusion that the bodies ought to accelerate differently?

Newton, of course, not only incorporates Galileo's well-established and widely recognized, if surprising, fact about bodies into his physical theory, he can also provide an explanation for Galileo's fact that is unavailable to the mechanists, and even to Galileo himself. He can indicate why the acceleration of any body toward earth is independent of the body's mass – it depends only on the earth's mass – and is therefore the same for all bodies. Since the force on any body, A, is equal to the mass of A multiplied by A's acceleration, we can substitute $m_A a_A$ into the law of gravitation, as follows. Let the mass of earth be given by m_B, and the mass of any body falling freely toward earth by m_A; then their gravitational interaction is measured by:

$$F_{grav} = G \; m_A m_B / r^2$$

where r is the distance between A and B and G is a constant.[79] Since the F_{grav} on A is given by $m_A a_A$, we can substitute this into the law to obtain the acceleration on A: F_{grav} on $A = m_A a_A$. But then $a_A = G \, m_B / r^2$, since m_A cancels out. Therefore, the acceleration earth produces on any body is dependent solely on earth's mass, and completely independent of the mass of the body. Since the mechanists lack mass, they cannot explain this phenomenon. For these reasons Newton would presumably regard his

---

[77] To my knowledge, Locke never discusses free fall. As for his understanding of heaviness, he indicates (*Essay* 4.6.11; cf. 2.31.9–10 and 3.6.30ff.), that weight – and therefore heaviness – are not primary qualities because they are not intrinsic, nor are they essential, properties of bodies. That is apparently one reason why he lists weight along with color as a property that would vanish given certain possible relations, or changes in relations, to other objects. Leibniz provides his own explanation of free fall, but I cannot delve into it here. Thanks to Karen Detlefsen for pointing this out.

[78] See book III, proposition 6 (*Principia*, 806) and the Scholium to the laws of motion (*Principia*, 424) for relevant discussions.

[79] As I indicated in ch. 3 above, this is our formulation of the law, and not Newton's own. For my purposes here, however, the anachronism is harmless.

conception of matter as superior to the mechanist conception, and it seems reasonable to endorse that assessment.

In these respects, Newton's concept of matter expresses a fundamental rejection of the prevailing mechanist conception of his day, just as his treatment of force in the *Principia* involves such a rejection. We might even call this rejection a revolutionary move within natural philosophy in the late seventeenth and early eighteenth centuries.

5

Space in physics and metaphysics: contra *Descartes*

When we think of seventeenth-century conceptions of space, time, and motion, we do not tend to think of Descartes as making a crucial contribution; and when we think of Descartes's philosophy, we do not tend to think of his understanding of space, time, and motion as particularly significant. But Descartes's views – as part of what has aptly been called his metaphysical physics – are essential for understanding Newton's conception of space and motion. The publication by Marie Boas and Rupert Hall of Newton's previously unpublished manuscript, now known as *De Gravitatione*, in the early 1960s helped to underscore this fact, for the text provides an extensive response to Descartes's view of space and motion in *Principia Philosophiae*.[1] Indeed, *De Gravitatione* greatly clarifies Newton's more famous, but much more concise, discussion of space, time, and motion in the Scholium to the *Principia*, first published some forty-three years after Descartes's work appeared. The discussion in *De Gravitatione* of the failures of the Cartesian physics of motion sheds light on the motivations for introducing absolute space in the Scholium to the *Principia*. It also clarifies and expands the characterization of God's relation to space in the General Scholium that lies at the heart of Newton's divine metaphysics.

Nonetheless, there are crucial differences between *De Gravitatione* and the *Principia*. For instance, although the former clarifies the discussion of

---

[1] See especially Garber, *Descartes's Metaphysical Physics*, for discussion. In *Unpublished Scientific Papers of Isaac Newton*, Hall and Hall provide a transcription of the original Latin of *De Gravitatione*. As they explain, the provenance of *De Gravitatione* remains unclear: the text bears no title – it is now known simply from its first line – and no date, although it is undoubtedly one of Newton's manuscripts. Despite extensive scholarly debate, there is no consensus on the proper dating of the text; Feingold has suggested that Newton may have composed different pieces of the manuscript at different times, but the question of dating will not be relevant to my discussion here (see *The Newtonian Moment*, 25–6). On Newton's relation to Descartes in general, see especially Koyré, *Newtonian Studies*, Böhme, "Philosophische Grundlagen der Newtonschen Mechanik," 11–13, and Cohen, "Newton and Descartes."

absolute space in the latter, *De Gravitatione* never addresses the famous distinction between relative and absolute motion. More significantly there is an aspect of *De Gravitatione* that casts doubt on the absolutism famously presented in the Scholium.[2] In *De Gravitatione*, after providing an extensive discussion of the Cartesian view of space and motion in *Principia Philosophiae*, Newton shifts his attention to the ontology of space, writing:

Perhaps now it may be expected that I should define extension as substance, or accident, or else nothing at all. But by no means, for it has its own manner of existing which is proper to it and which fits neither substances nor accidents. It is not substance: on the one hand, because it is not absolute in itself, but is as it were an emanative effect of God and an affection of every kind of being [*non absolute per se, sed tanquam Dei effectus emanativus, et omnis enties affectio quaedam subsistit*]; on the other hand, because it is not among the proper affections that denote substance, namely actions, such as thoughts in the mind and motions in body. (*De Gravitatione*, 21; Hall and Hall, *Unpublished Scientific Papers*, 99)

However, in the Scholium to the *Principia*, Newton obviously treats space as absolute, making this the centerpiece of his discussion. Could a text that sheds so much light on the Scholium conflict with it on such a basic issue?

In this chapter I argue that these views can in fact be reconciled with one another. In the course of reconciling them, one recognizes that Newton carefully separated various questions about space – some of these questions must be addressed at the beginning of a treatise in physics, where certain presuppositions about the physical world are appropriate, while others must await a more general metaphysical discussion, where such presuppositions are questioned, and where some of the known physical facts are bracketed. Interpreting Newton's various claims about space, therefore, helps to illuminate his overarching understanding of the relation between physics and metaphysics. That understanding, as is well known, differs fundamentally from the conception one finds in Descartes. But just as significantly, both Newton's physics and his metaphysics are intended to rebut Cartesian views.

---

[2] It seems that Newton *may* have defended a version of absolutism early in his career – for instance, in the so-called *Wastebook* from (roughly) 1666 (see Hall and Hall, *Unpublished Scientific Papers of Isaac Newton*, 157).

SPACE AND THE LAWS OF MOTION

In *Principia Philosophiae,* Descartes distinguishes the "vulgar" from the "proper" conceptions of motion, where the proper, or strict, conception is classically relationalist.[3] After discussing motion in the vulgar or ordinary sense in section 24 of part II of *Principia Philosophiae,* Descartes writes in section 25:

> But if we consider what should be understood by *motion,* not in common usage but in accord with the truth of things, and if our aim is to assign a determinate nature to it, we may say that *motion is the transfer of one piece of matter or one body from the vicinity of the other bodies which immediately touch it, and which we consider to be at rest, to the vicinity of others* [*ex vicinia eorum corporum, quae illud immediate contingent & tanquam quiescentia spectantur, in viciniam aliorum* | *du voisinage de ceux qui le touchent immediatement, et que nous considerons comme un repos, dans le voisinage de quelques autres*]. (VIII-1: 53–4, IX-2: 76)

This distinction between the "vulgar" and the "proper" conception of motion may be demanded by Descartes's metaphysics. The *vulgare* conception of motion – where motion consists in "the action by which a body travels from one place to another" – may reflect our ordinary ideas about motion, but if one thinks that extension is the essence of body, and that there cannot be any empty space, then wherever there is a place, there is a body (or bodies). Any "travel" from place to place will necessarily involve a change in the traveling body's relation to other bodies. And any body will necessarily be surrounded by other bodies – its "vicinity" – at every instant of its existence. The *propria* view of motion reflects these views, jettisoning the notion of a place.

Many of Newton's objections to *Principia Philosophiae* in *De Gravitatione* reflect an overarching attitude toward Cartesian proper motion, viz. that it fails to reflect facts about motion expressed by Descartes's own laws of nature. Descartes's first two laws of nature in *Principia Philosophiae* are introduced as follows (part II, sections 37–9):

---

[3] Here I refer only to a view of *motion,* and not to a relationalist conception of *space à la* Leibniz. The distinction between a vulgar and a proper (or mathematical) conception of space, time, or motion was obviously known to Newton through *Principia Philosophiae.* But he may also have encountered it in Isaac Barrow's work – see, for instance, the tenth of his mathematical lectures in Barrow, *The Usefulness of Mathematical Learning,* 163–4. Hall discusses Barrow's influence on Newton in "Newton and the Absolutes," 273, 278–9. On Descartes's conception of motion, see Garber, *Descartes's Metaphysical Physics,* 160–88.

The first law of nature [*lex naturae*]: that each and every thing, in so far as it can [*quantum in se est*], always perseveres in the same state, and thus what once moves always continues to move . . . The second law of nature: that all motion is in itself rectilinear and hence any body moving in a circle tends always to recede from the center of the circle it describes. (VIII-I: 62–3, IX-2: 84–5)[4]

Newton claims, for instance, that Descartes is inconsistent regarding the ever-important issue of the earth's motion. On the one hand, Descartes's conception of "proper" motion implies that a body B moves if two conditions are met: first, B is transferred from one group of surrounding bodies to another group; and second, B's original group of surrounding bodies is regarded as not moving. Since the earth is carried around its solar orbit by a vortex that entirely surrounds it, Descartes concludes that the earth does not move, properly speaking. On the other hand, Descartes says that the earth has a tendency to recede from the sun (Newton cites *Principia Philosophiae*, part III: section 140). But by Descartes's laws, if the earth were at rest, it would remain at rest and would *not* tend to recede from the sun – it would only have such a tendency if it were following a curvilinear trajectory.[5]

According to Newton (*De Gravitatione*, 15–16), Descartes's view that each body has only one "proper" motion also conflicts with his definition of such motion.[6] Imagine two observers in separate spacecraft watching the earth and its vortex flow through the solar system (imagine, for the sake of argument, that the vortex is perceptible).[7] One observer maintains an unchanging position external to the vortex surrounding the earth, and regards the vortex as being at rest. From her point of view, if the earth remains surrounded by the vortex, it does not move, properly speaking; but if the earth is transferred away from this vortex, it does move, since she regards the vortex as being at rest. If a second observer is placed so that he regards the vortex surrounding the earth at time$_1$ as moving, then given the

---

[4] Descartes is probably the first seventeenth-century author to consider rectilinear motion the "natural" state of motion for a body. Galileo, and even Gassendi, apparently took circular motion to be a natural state under certain circumstances – see Bertoloni Meli, *Thinking with Objects*, 145–7 for an illuminating discussion. On Gassendi in particular, see LoLordo, *Pierre Gassendi*, 177–8. Huygens gives his formulation of the principle of inertia in part II, hypothesis I of his *Horologium* of 1673 (*The Pendulum Clock*, 33).

[5] Cf. the discussion in *Principia*, 413–14.

[6] See *Principia Philosophiae*, part II: section 31; Newton, in *De Gravitatione*, 14, also cites sections 28 and 32 (which correspond to VIII-I: 57, 55 and 57 respectively).

[7] It is not essential to Newton's objection here that the earth be surrounded by a vortex, for the objection would hold if it were surrounded by some other body or bodies and, of course, Descartes is committed to the view that the earth must be surrounded by some body or bodies.

definition above, he cannot regard the earth as moving, even if it is transferred to the vicinity of other bodies at time$_2$. The reason is that the earth
cannot move, beginning at time$_1$, if at time$_1$ the vortex surrounding it is not
regarded as at rest. So these observers will disagree if they each take the earth
to be transferred away from the vicinity of the vortex that (they agree)
surrounds it at time$_1$.

Newton also objects to the fact that Descartes renders a body's proper
motion relative to its position with respect to other bodies. Newton makes
his case as follows: suppose that the vortex surrounding the earth were
moving according to Descartes's view of proper motion – i.e. the vortex is
transferred from the bodies surrounding it, which we regard as being at rest.
This means the earth must be at rest. If God were to render the vortex
surrounding the earth motionless, without interacting with the earth in any
way, then a formerly stationary earth would begin moving (as long as we
regarded the vortex as being at rest). God could therefore move the earth
without applying a force to it, or interacting with it in any way. Once again,
we find a tension with Descartes's laws, since the first law indicates that a
body at rest will remain at rest unless acted upon – to explain the first law,
Descartes writes: "If it is at rest, we hold that it will never begin to move
unless it is pushed into motion by some cause" (*Principia Philosophiae*,
part II: section 37). From Newton's point of view, it is a mistake to sever the
tie between true motion and external action, a tie Newton emphasizes in the
Scholium to the *Principia*, as we will see.

Notice that none of these arguments concerns the debate between
relationalist and absolutist conceptions of space; that is, the objections
are to Descartes's conception of proper motion, and not to the overarching view that space is the order of actual and possible object relations,
a view one finds prominently in Leibniz's letters to Clarke, but not
in *Principia Philosophiae*. This is significant because the arguments in
De Gravitatione shed light on Newton's discussion in the Scholium, which
is often interpreted primarily through the lens of the relationalism–
absolutism debate.

The Scholium follows the definitions at the beginning of the *Principia*,
before book I begins; Newton notes that he will not define space, time and
motion, as he did such quantities as centripetal force, mass, and the quantity
of motion:

Thus far it has seemed best to explain the senses in which less familiar words are
to be taken in this treatise. Although time, space, place, and motion are very
familiar to everyone, it must be noted that these quantities are popularly
conceived solely with reference to the objects of sense perception. And this is

the source of certain preconceptions; to eliminate them it is useful to distinguish these quantities into absolute and relative, true and apparent, mathematical and common. (*Principia*, 408)[8]

Although Newton thinks the common conception of space, time, and motion leads to problematic "preconceptions," the fact that he does not define space, time, and motion here is crucial, for he begins with the common understanding of these quantities, and then transforms that conception into a mathematical understanding of them. Hence Newton writes that absolute and relative space are the same "in species and magnitude"; that is, both absolute and relative space are Euclidean magnitudes (three-dimensional magnitudes with Euclidean structures).[9] A space is relative just in case it is a Euclidean magnitude whose parameters are defined by (perceptible) object relations. Since absolute space is the same type of quantity, we can achieve a conception of it by a simple procedure. To conceive of absolute space, we would conceive of a relative space, and then remove its parameters by excising the objects and relations inhabiting the relative space – with no parameters, the space expands infinitely in all directions; its infinity entails its immobility. Since both relative and absolute space are homogeneous, no property of absolute space is left out in this procedure. For absolute space is an infinite, immobile, homogeneous, Euclidean magnitude.

   After distinguishing absolute and relative space Newton distinguishes between absolute (or true) and merely relative motion: "Absolute motion is the change of position of a body from one absolute place to another; relative motion is change of position from one relative place to another" (*Principia*, 409). But why do we require absolute motion? Why is it not sufficient to think of each body's motion as involving changes in its relations to other

---

[8] Although Newton does not define space, time, and motion here, he makes it clear from the outset that he takes each of these to be a "quantity." This is appropriate: in a sense, the *Principia* as a whole focuses on quantities that can be measured, which includes forces of various kinds, mass, acceleration, and even space, time, and motion – see ch. 3. For an illuminating discussion of the Scholium, see Rynasiewicz, "By Their Properties, Causes and Effects: Newton's Scholium on Time, Space, Place and Motion," which dispels several common misunderstandings of the text.

[9] A relative space for Newton can perhaps be arbitrarily large, but presumably must be finite because it is "determined by our senses from the situation of the space with respect to bodies" (*Principia*, 409). A relative space could presumably not be constituted by a body that was infinitely distant from the perceiver defining the space, since that would render it imperceptible. This may indicate that relative spaces are Euclidean magnitudes only in a restricted sense, since Euclidean geometry might be taken to imply that we can, for instance, draw a ray from any arbitrarily chosen point and continue it infinitely in one direction. Since Newton himself focuses on other properties of Euclidean space, such as homogeneity, which would hold even for a finite space, I leave this issue aside in what follows. Thanks to Michael Friedman for clarifying this point.

bodies, or perhaps to relative places that are defined in terms of such relations?

Newton provides us with at least four reasons to jettison the view that we can understand a body's true motion in terms of changes in its relations to other bodies (Descartes gives one construal of this overarching view). First, he notes an empirical fact, namely that there may be no body that is truly at rest anywhere in the universe to which we could refer the relative motions of all other bodies (*Principia*, 411). Second, he notes that although there may in fact be a body that is truly at rest, it is perfectly possible that we will be unable to determine which body is at rest, and that some other body within the reach of our senses – say within our solar system, if we count astronomical observation here – maintains a fixed position with respect to the truly resting body. Third, in a point that echoes *De Gravitatione*, Newton underscores a difficulty that may be specific to the Cartesian construal of proper motion. We would typically say that if a body's parts maintained their positions relative to one another, then if the body itself moved, its parts would move as well. For instance, if I throw a peanut into the air, we would ordinarily say that the peanut inside moves along with its shell, despite the fact that the peanut and the shell maintain their positions with respect to one another. But this means that we cannot define the true motion of a body as involving its transference away from its "vicinity" *à la* Descartes.[10]

With his fourth point, Newton shifts gear, emphasizing that if we take true motion to involve a change in object relations, we will sever the tie between motion and its causes:

> The causes which distinguish true motions from relative motions are the forces impressed upon bodies to generate motion. True motion is neither generated nor changed except by forces impressed upon the moving body itself, but relative motion can be generated and changed without the impression of forces upon this body. (*Principia*, 412)

Notice that this understanding of the relationship between true motion and impressed force is independent of the view that true motion is motion with

---

[10] I. B. Cohen helpfully indicates that Newton's language in the Scholium closely parallels the language Descartes uses in his *Principia Philosophiae* ("Newton and Descartes," 620). Newton writes: "true and absolute motion cannot be determined by means of change of position from the vicinity of bodies that are regarded as being at rest [*motus verus & absolutus definiri nequit per translationem e vicinia corporum, quae tanquam quiescentia spectantur*]" – *Principia*, 411; *Principia Mathematica*, vol. I: 49. As noted above, Descartes defines proper motion as follows: "motion is the transfer of one piece of matter or one body from the vicinity of the other bodies which immediately touch it, and which we consider to be at rest, to the vicinity of others [*ex vicinia eorum corporum, quae illud immediate contingent & tanquam quiescentia spectantur, in viciniam aliorum*]" – *Principia Philosophiae*, VIII-1: 53.

respect to absolute space. The idea is that a body's relations to other bodies do not bear this relationship to impressed forces, for as Newton indicates in *De Gravitatione*, we can alter the relations of a body without impressing any force on it; and even if we impress a force on a given body, if we impress forces on the bodies that bear a relation to it, their relations might remain unchanged. Therefore, in order to understand the relation between true motion and impressed force, we should not understand true motion in terms of a body's relations with other bodies. This helps to place Newton's move to absolute space in the right light: in order to understand the relationship between true motion and impressed force, we construe true motion in terms of a body's relation to places within absolute space, rather than in terms of its relation to other bodies.

The treatment of the famous rotating bucket can profitably be understood within this context. First, Newton notes that acceleration (e.g. true rotation) is empirically detectable through the presence of inertial effects, even in the absence of a change in object relations. Second, Newton contends, *contra* Descartes, that we cannot understand the true motion of the water in the bucket as consisting of changes in relations between the water and a surrounding body (in this case, the bucket). The relation between the water and the bucket remains the same, despite the fact that the water has true motion, as indicated by the presence of inertial effects. So the true motion of a body cannot be understood in terms of changes in its relations to other objects. Absolute space allows us to capture what true motion is.

If we interpret the bucket example within the context of a discussion of Cartesian proper motion, rather than Leibnizian relationalism, an important consequence follows. To show, *contra* Descartes, that motion in the true sense is not to be understood in terms of a body's relation to those bodies that are immediately contiguous to it, Newton can ignore the question of whether other entities exist in the world of the spinning bucket. The world can be populated arbitrarily by however many entities one likes, for the bucket immediately surrounds the water, and is therefore the only relevant object – according to Descartes – for assessing the water's true motion. If Newton were in fact attempting to undermine Leibnizian relationalism, the question of whether other objects exist would become crucial. For Leibniz would say that the water truly moved if it underwent an observable change with some other object – the vicinity of a given body would drop out as irrelevant.[11] And certainly if other objects exist, it is not

---

[11] See especially Leibniz's fifth letter to Clarke, sections 52–3 (*Die philosophischen Schriften*, vol. VII: 403–4).

difficult to conceive of circumstances in which we could, at least in principle, observe the water to be undergoing changes in its relations to other
objects; perhaps not to the bucket but to the distant fixed stars. So Newton's
discussion may successfully undermine Descartes's view without touching
Leibniz's relationalism.

The bucket experiment also highlights the importance of distinguishing
between our ordinary understanding of when an object is moving, and what
Newton himself calls the "vulgar" or ordinary conception of motion. We
would ordinarily say that the water in the bucket was moving when its
surface was concave – we do not require any technical notions or arguments
to reach this conclusion. Newton then enlists that conclusion to support his
understanding of true motion. This subtlety was apparently lost on later
empiricist critics of the bucket example. For instance, Berkeley is committed to the view that all motion is relative, presumably on the grounds that
this reflects common sense. But this places him in a bind. In his response to
Newton's bucket example in *Principles of Human Knowledge*, Berkeley must
stipulate that when the water reaches what we would ordinarily take to be
the height of its motion – when it is most concave – it is in fact at rest
because its situation with respect to the bucket is no longer changing.[12] It
seems that Berkeley's attempt to defend common sense against the metaphysical extravagances of Newtonian space fails on this score. Newton's
move here is clever: the absolute view of motion does not conflict with, but
rather employs, our ordinary understanding of the water's motion.

Ironically enough, Newton's contention that Descartes fails to present a
conception of space, time, and motion that reflects the laws of nature
presented in *Principia Philosophiae* has an echo in Newton's own work,
although it was one that was not articulated until long after Newton's death.
Broadly stated, there is a significant tension between Newton's own conception of true space and motion on the one hand, and his three laws of
motion, along with their corollaries, on the other. The notion of absolute
space implies that each body has a velocity relative to space at any given
instant. But the laws of motion indicate that the forces acting on a body are

[12] Berkeley introduces Newton's *Principia* in section 110 of *Principles of Human Knowledge* – calling it "a
certain celebrated Treatise of *Mechanics*" – and then notes in section 112 that all motion must be
relative. Citing this discussion, he then writes in section 114: "For the water in the vessel, at that time
wherein it is said [by Newton] to have the greatest relative [sic] circular motion, hath, I think, no
motion at all: as is plain from the foregoing section." This conclusion presents Berkeley with a
difficulty: he takes the water to be moving as long as its relation to the bucket changes, and then when
that relation ceases to change, it is at rest; therefore, there ought to be a dynamic effect associated with
this sudden stopping of the water's motion – for instance, it ought to slosh around when it stops – but
of course there is no such effect.

independent of its velocity – hence two bodies bumping into each other exert forces on one another that are proportional to their accelerations (and masses), but this is independent of whether these bodies are at absolute rest or are moving inertially with any velocity whatever. As Newton himself notes, the laws of motion imply corollary five, which indicates that no experiment could determine whether any closed system of bodies was at absolute rest or moving inertially – for all the forces and accelerations are independent of velocity in absolute space (*Principia*, 423). But this indicates that the notion of absolute space gives rise to a quantity – the true velocity of each object – that can never be measured. Moreover, it indicates that Newton defines true motion not in terms of his own laws, but rather in terms of changes in absolute place.

Nonetheless, this criticism of absolute space – and the subsequent development of the notion of an inertial frame, which replaces absolute space and leaves us with quantities that can be given an empirical meaning, unlike true velocity – was never broached in Newton's lifetime. Indeed, his primary interlocutors and critics, especially Leibniz, Huygens, and Berkeley, focused instead on various metaphysical objections to absolute space and time. That is, Newton's philosophical critics did not meet Newton on his own ground – they did not determine what conception of space, time, and motion was presupposed by, or fitted most closely with, Newton's own laws of motion. On the contrary they ignored this issue, electing instead to provide independent philosophical objections – based on the principal tenets of classical empiricism, in Berkeley's case, or the principle of sufficient reason, in Leibniz's case – to Newton's conception of absolute space, time, and motion. In that regard, they missed an important aspect of Newton's own approach, one expressed in his criticisms of Descartes.

Once I have outlined Newton's discussion of the ontology of space in *De Gravitatione*, I will be in a position to consider the coherence of Newton's views of space within both physical and metaphysical contexts in the following section below.

THE ONTOLOGY OF SPACE

Newton's discussion in *De Gravitatione* is not limited to highlighting the failures of Cartesian physics. Once he finishes his extensive rejection of the Cartesian conception of motion, he shifts to a more general discussion of the ontology of space. But his discussion is clearly still informed by his reading of Descartes, this time by his understanding of Cartesian metaphysics. He begins his discussion of space by connecting it to Cartesian dualism:

For since the distinction of substances into thinking and extended, or rather into thoughts and extensions, is the principal foundation of Cartesian philosophy, which he contends to be known more exactly than mathematical demonstrations: I consider it most important to overthrow [that philosophy] as regards extension, in order to lay truer foundations of the mechanical sciences. (*De Gravitatione*, 21)

The very next paragraph considers what we might think of as a Cartesian presupposition, viz. the view that any item within our ontology must be considered either a substance in its own right, or else an accident (or a mode) of some substance. He writes:

Perhaps now it may be expected that I should define extension as substance, or accident, or else nothing at all. But by no means, for it has its own manner of existing which is proper to it and which fits neither substances nor accidents. It is not substance: on the one hand, because it is not absolute in itself, but is as it were an emanative effect of God and an affection of every kind of being; on the other hand, because it is not among the proper affections that denote substance, namely actions, such as thoughts in the mind and motions in body. (*De Gravitatione*, 21)[13]

Newton explicitly introduces the claim that space is an "affection" – a claim I will call "the affection thesis" – as a response to a question about space that a Cartesian would find pressing, and in contending that space is neither a substance nor an accident, but rather something called an affection, Newton obviously intends to depart from a Cartesian metaphysical framework. But precisely how does an affection differ from a substance or a property?

To make progress in understanding Newton's view, we have to distinguish three concepts of substance found in his work. This will not exhaust Newton's understanding of substance, but it should be sufficient to comprehend his claims concerning the status of space.[14] First, there is what some moderns pejoratively regarded as an unintelligible Scholastic conception, according to which a substance is characterized by a "substantial form." Since Newton rejects this idea *per se* there are no substances in this sense (*De Gravitatione*, 29), which renders this concept irrelevant in this context, for then it would be trivial to deny that space was a substance.

---

[13] Newton may have been influenced by Gassendi – whose work he knew through Walter Charleton's *Physiologia Epicuro-Gassendo-Charltoniana* – when developing this view; for discussion, see LoLordo, *Pierre Gassendi*, 106–24. Newton may also have been influenced by Barrow, who contended that space seemed to be neither an accident, nor a self-subsisting thing, in the tenth lecture included in *The Usefulness of Mathematical Learning*, 164. However, Barrow appears to use "affection" to mean property, which differs from Newton's usage of that term in his theory of space in *De Gravitatione* (cf. Barrow, *The Usefulness of Mathematical Learning*, 182–4). Or so I will argue.

[14] Thanks to Justin Broackes for discussion of this point. For a nice discussion of the notion of substance in various historical epochs, see his "Substance."

Second, a substance can be conceived of as the subject of actions, where each substance-type exhibits a characteristic kind of action: material bodies move, minds think, etc. (*De Gravitatione*, 33). Newton was perfectly well aware that this criterion for substancehood was not standard:

[A]lthough philosophers do not define substance as an entity that can act upon things, yet everyone tacitly understands this of substances, as follows from the fact that they would readily allow extension to be substance in the manner of body if only it were capable of motion and of sharing in the actions of body. And on the contrary, they would hardly allow that body is substance if it could not move, nor excite any sensation or perception in any mind whatsoever. (*De Gravitatione*, 21–2)

Although his criterion is not widely articulated, then, he thinks that it would be accepted as reasonable by his readers. Newton obviously intends action to be broadly construed in the context of his criterion, since he takes the mind to be a substance, with thought as its characteristic action.[15] Space is non-trivially not substantial in this sense, since it is causally inert and therefore exhibits no actions. This indicates that space, unlike objects and minds, is not a substance in a clear and straightforward sense.

Finally, there is a third concept according to which neither objects, nor minds, nor space is a substance – we might call this the "Cartesian" concept of a thing that exists independently of everything else that exists.[16] Only God meets this definition of a substance, since objects and minds are contingent beings and are therefore dependent on God. Space, too, is not a substance in this sense, for it depends on God for its existence. Or so I will argue.

After discussing Descartes's understanding of space in more depth, especially his thoughts about the possible infinity of space, Newton returns to the affection thesis, clarifying it as follows:

4. Space is an affection of a being just as a being. No being exists or can exist which is not related to space in some way. God is everywhere, created minds are somewhere, and body is in the space that it occupies; and whatever is neither every where nor anywhere does not exist. And hence it follows that space is an

---

[15] In a draft of a potential new preface to the *Principia* written sometime after 1713, Newton discussed the actions of our mind of which we are conscious – see *Principia*, 53 and 54. As I discuss below, although substances can be sorted into types by their characteristic actions, there is another sense in which all of these substance-types fall under a single broader genus, viz. that of extended substance, for Newton regards all substances as extended.

[16] See *Principia Philosophiae*, VIII-1: 24. Foreseeing this difficulty, the influential Cartesian natural philosopher Rohault defines a substance as a "thing which we conceive to subsist of itself, independent of any other created thing" – see Rohault, *System of Natural Philosophy*, vol. I: 15. This is the English translation of Rohault's important textbook, which in later editions included Samuel Clarke's avowedly "Newtonian" notes on the Cartesian book.

emanative effect of the first existing being, for if any being is posited, space is posited. (*De Gravitatione*, 25)

This is a clear rejection of Cartesian dualism. For Newton, the Cartesian *res cogitans* is nowhere – it is not extended – and therefore does not exist.[17] Newton explicitly regards all entities as spatial, including human minds and even God, and thereby rejects Cartesian dualism. In a crucial sense, there is only one kind of substance. At least, from a Cartesian point of view, this is one way of characterizing Newton's thought.[18]

It seems to me that Newton's view in the passage just quoted has the following structure:[19] (1) space is an affection of every kind of being; (2) this is clarified by the claim that God is spatially ubiquitous,[20] the mind is spatial, and bodies occupy certain spaces, and also by the claim that an entity that is nowhere in space does not exist; (3) the affection thesis entails the claim that space is an "emanative effect" of the first existing being; and (4) the emanation thesis is clarified by the claim that if any being is posited, space is posited.

But the affection thesis remains unclear. If the thesis asserts that space depends on other entities for its existence – and is therefore not a substance in an important sense – then one wonders why space is not simply a property. I take it that if space were a property, it would not only depend for its existence on its bearer, it would presumably have characteristics that depended on its bearer. For instance, a finite substance presumably cannot have certain infinite properties – it could not be infinitely long, or infinitely heavy, etc. – so the character of the bearer would constrain the character of the property.[21] But that evidently is not the case with the affection that

---

[17] Newton defends a similar view in an unpublished manuscript, "Tempus et Locus," which probably originates in the 1690s, after the first edition of the *Principia* appeared: "Time and place are common affections of all things [*Tempus et Locus sunt omnium rerum affectiones communes*] without which nothing whatsoever can exist. All things are in time as regards duration of existence, and in place as regards amplitude of presence. And what is never and nowhere is not in rerum natura" – McGuire, "Newton on Place, Time and God," 117. McGuire first transcribed and published the manuscript in 1978; I use his translation.

[18] Thanks to Daniel Garber for discussion of this issue.

[19] In what follows, I defend an interpretation of Newton's view that parallels the interpretation in Stein, "Newton's Metaphysics." However, as I outline below, I differ from Stein on other matters.

[20] Newton would certainly have been aware that his understanding of God's spatial ubiquity was not unique in this period. We find such a conception, for instance, in Barrow (*The Usefulness of Mathematical Learning*, 170–1, 178) and in Henry More (appendix to *An Antidote Against Atheism*, 334ff.; cf. also *Immortality of the Soul*, 1.4.3 and 1.4.4). Intriguingly, both Barrow and More contemplate counterfactuals regarding the status of space in a divinely inhabited world with no, or with spatially restricted, matter.

[21] Perhaps there are some infinite properties, such as infinite divisibility, that finite substances can in fact bear. But other criteria distinguish affections from properties, so I leave this issue aside here. Thanks to Tad Schmaltz for making this point.

Newton has in mind, for Newton explicitly takes space to be infinite, but takes it to be an affection of finite entities, such as ordinary physical objects. Unlike properties and modes the character of the first existing entity in no way constrains the character of the affection, in this case space. An affection's existence is dependent, but its characteristics are independent *per se*.

More definitively space *qua* affection is not dependent on any given object, as would be the case for modes and properties.[22] Properties come and go with their bearers, as do modes, but an affection is not dependent on any particular thing (unless of course that thing is lonely in its world). Instead, it exists just in case any entity exists. Hence in a universe in which one frog exists, and then disappears precisely as another frog appears, and so on, space would exist continuously. Indeed, space and its characteristics are invariant under all possible causal interactions among objects – which is obviously not the case with any property or mode, for those come and go as their bearers interact causally with other things – with the sole exception of the disappearance of all entities. That is the only natural change that could affect space (assuming God's non-existence – see below).

What about the other components of Newton's view? He says that the affection thesis entails that "space is an emanative effect of the first existing being." What precisely does that mean? As Newton well knew, Henry More

---

[22] In a passage from a preliminary draft to the *Avertissement au Lecteur* that Newton wrote for Des Maizeaux's edition of the Leibniz–Clarke correspondence, he clarifies the denial that space is a property:

The reader is desired to observe, that wherever in the following papers through unavoidable narrowness of language, infinite space or immensity and endless duration or eternity, are spoken of as *qualities* or *properties* of the substance which is immense or eternal, the terms *quality & property* are not taken in that sense wherein they are vulgarly, by the writers of *logick & metaphysicks* applied to *matter;* but in such a sense as only implies them to be modes of existence in all beings, & unbounded *modes* & consequences of the existence of a substance which is really necessarily and substantially omnipresent and eternal: which existence is neither a substance nor a quality, but the existence of a substance with all its attributes properties & qualities, & yet is so modified by place & duration that those modes cannot be rejected without rejecting the existence. (Cohen and Koyré, "Newton and the Leibniz–Clarke Correspondence," 96–7)

What Newton calls here a "mode" of existence is termed an "affection" in *De Gravitatione*; just as he asserts in *De Gravitatione* that space and time are affections of every kind of being, here he takes them to be "modes of existence in all beings" (*ibid.*, 97). And just as he claims here that space and time are consequences of the existence of God, in *De Gravitatione* he contends that space and time are "emanative effects" of God. It is also quite clear in this passage, incidentally, that Newton avoids Clarke's claim in his correspondence with Leibniz that space is a property of God (C 3: 3, C 4: 10, *Die philosophischen Schriften*, vol. VII: 368, 383). Perhaps Clarke was attempting, in his own way, to deal with what Newton calls the "unavoidable narrowness of language" in this area, but Newton's formulation may be more successful.

was famous for claiming that space was an "emanative effect" of God.[23] John Carriero provides a powerful discussion of the view that Newton follows More in his thinking about emanation.[24] From Carriero's point of view, Newton endorses the Morean conception that emanation involves efficient causation, so the claim is that God efficiently causes space and time, a position Clarke defended in his correspondence with Leibniz.[25] But Carriero also emphasizes a site of disagreement between Clarke and Newton – Clarke fails to appreciate fully Newton's emanation thesis, misrepresenting Newton's view by telling Leibniz that space is akin to one of God's properties.[26] Carriero ends his paper with the following criticism: although Newton contends that God's mere existence efficiently causes space and time to exist, he is saddled with the view that God cannot exist independently of space and time. I focus first on Carriero's understanding of emanation, saving his criticism of Newton for later.

Carriero claims that Newton follows More in thinking that space and time emanate from God, by which he means that God's mere existence efficiently causes space and time. I will argue that this interpretation is inconsistent with three aspects of Newton's view: (1) the fact that the affection thesis entails the emanation thesis; (2) the fact that Newton glosses

---

[23] More was an influential and powerful figure in Cambridge when Newton arrived at Trinity College in the 1660s; the two knew one another personally – having attended the same grammar school in Grantham, where one of Newton's teachers was a former student of More's – and Newton had several of More's texts in his personal library. See *Correspondence*, vol. II: 306 and Westfall, *Never at Rest*, 97.

[24] Although Carriero responded to a paper of McGuire's, I take him to have provided a stand-alone interpretation of Newton's conception of space. See McGuire, "Predicates of Pure Existence: Newton on God's Space and Time," and Carriero, "Newton on Space and Time: Comments on J. E. McGuire." In some respects, my view is closer to McGuire's original view, before he conceded certain points to Carriero.

[25] Although he does not address the relation between More's and Newton's views, and although he does not make use of the notion of emanation, Clarke contends in his fifth letter to Leibniz that God's "existence causes space and time" (C 5: 36–48; *Die philosophischen Schriften*, vol. VII: 427). But there is no consensus on these issues: McGuire (*Tradition and Innovation*, 15) denies that Newton and More have the same understanding of emanation, as does Stein in "Newton's Metaphysics." In other respects, however, I differ from both McGuire and Stein, as will be evident below.

[26] Carriero makes three principal points concerning *De Gravitatione*: first, we must distinguish between Newton's discussion of the "real beings" space and time, and his discussion of the "affections" of objects, which Carriero describes as the spatial and temporal locatability of objects; second, Newton follows More in his understanding of emanation; and third, in contending that space and time are efficiently caused by God, Newton is committed to the denial of Clarke's view that space and time are properties of God, for that would render God their material, rather than their efficient, cause. Carriero's claim that Newton distinguishes between space and time *qua* "real beings," and space and time *qua* "affections," may conflict with central passages in *De Gravitatione*. In the passage quoted above, for instance, Newton denies that space is a substance or an accident, asserting instead that it "has its own manner of existing," for it is "an emanative effect of God and an affection of every kind of being." Hence Newton thinks that space, the "real being," is in fact an affection, and an emanation from God. But I leave this point aside in what follows.

the emanation thesis with his claim about "positing"; and (3) the fact that the emanation thesis holds not for God *per se* but for what Newton calls "the first existing being." Let us take these in turn.

As for (1), what can the emanation thesis mean, given that Newton takes it to be entailed by the affection thesis? The latter is purely general: it is not the assertion that space is an affection of God, or of any particular type of entity; on the contrary, Newton takes space to be an affection of all entities. This is quite clear from his gloss on the affection thesis: God, the human mind, and all bodies occupy space in some way, which is taken to explain the claim that space is an affection of every kind of being. But it is unclear how this thesis about the relation between space and all bodies could be said to entail a thesis that concerns God alone. That is, Carriero's view must hold exclusively for God – presumably no finite entity could efficiently cause infinite space to exist. But it is difficult to see how the claim that all kinds of entity bear some relation to space could be said to entail the claim that space was efficiently caused by God.

As for (2), Newton clarifies the claim about emanation by saying that if any being is posited, space is posited – once again, as with the affection thesis, this is not a claim specifically about God, but rather about any entity whatsoever. Had Newton meant to be glossing a claim about God's relation to space, he presumably would have written that if we posit God, we posit space. In addition, the scope of Newton's claim about positing indicates that Newton would have to view space as efficiently caused by every kind of being, even the finite ones, which seems implausible.

Finally, as for (3): Newton does not claim that space emanates from God, but rather that it is an emanative effect of the first existing being, and as we have seen, this is entailed by the thesis that space is an affection of every kind of being. Had Newton intended to refer solely to God as the first existing being, he could have done so by saying that space emanates from a necessary being; instead, he left the description more general, leaving it open – at that stage in his discussion – what the first existing entity was, and what its characteristics were. Once we insert the additional premise that God is in fact a necessarily existing entity[27] – and indeed, the only such entity – we get the conclusion that space emanates specifically from God. But that additional premise is not part of the emanation thesis itself; nor does the

---

[27] Newton clearly took the view that God exists necessarily to be commonplace (*Principia*, 942) and therefore not in need of argumentative support – I discuss this issue in ch. 2 and again in ch. 6. In a text that Newton had in his personal library, *An Antidote Against Atheism* (1655), More also endorses that view – book I, ch. 4, 12ff.

affection thesis entail the claim that God exists necessarily. Instead, Newton's view is general: since space is an affection of every entity, it follows that space emanates from whatever entity is the first to exist. The generality of the claim precludes emanation from involving efficient causation, on pain of the suspicious conclusion that any entity – whether finite or infinite – efficiently causes space, just in case that entity is the first to exist.

This interpretation is confirmed, in turn, by Newton's discussion of the distinction between body and space. As usual, Newton intends to rebut Descartes's understanding of this distinction – in the course of doing so, he indicates how God's will relates to body and to space, especially in the context of his famous "creation story" concerning material objects:

And since there is no difference between the materials as regards their properties and nature, but only in the method by which God created one and the other, the distinction between body and extension is certainly brought to light from this. For extension is eternal, infinite, uncreated, uniform throughout, not in the least mobile, nor capable of inducing change of motion in bodies or change of thoughts in the mind; whereas body is opposite in every respect, at least if God did not please to create it always and everywhere. (*De Gravitatione*, 33)

When coupled with the points I have raised above, Newton's explicit claim that unlike body, space is "uncreated,"[28] may conflict with Carriero's reading. If Newton had asserted that space is created, then Carriero would have strong grounds for arguing that only God causes space to exist, since only God can be a creator. Since Newton denies that space is created, Carriero lacks a possible ground for thinking that Newton takes space to bear a special causal relationship with God. For his part, Carriero notes that the denial in this passage that space is created may merely amount to the denial that space is temporally posterior to its cause. But this point merely carries us back to the issues I have raised above: even if space is not temporally posterior to its cause, if Carriero were right overall, then the scope of Newton's views would entail that any finite entity could efficiently cause space to exist, while being simultaneous with space. And as I have argued, this reading of Newton is implausible.[29]

---

[28] Once again, Newton's view bears an important resemblance to the view defended by Gassendi – for discussion, see LoLordo, *Pierre Gassendi*, 121. LoLordo notes in passing some of the parallels between Newton's and Gassendi's views.

[29] For this reason, I think McGuire (in "Predicates of Pure Existence: Newton on God's Space and Time," 105; cf. McGuire, *Tradition and Innovation*, 15) should have stood his ground in response to Carriero's criticisms. McGuire originally denied that God causes space and time to exist, holding instead that they bear a kind of "ontic dependence" on God.

Of course, Carriero might respond by raising a distinct objection: if Newton does not follow More, what then causes space and time to exist, according to Newton? Surely they are not necessary beings, so their existence must have some cause.

However, if we consider the intersection of Newton's conception of God with his affection thesis – which entails the emanation thesis – we find that he actually thinks of space and time as uncaused. And indeed, as we have seen, Newton contrasts space with body by indicating that the former is "uncreated." But this does not follow from a general view of what Newton calls affections; it follows from God's specific relation to space. Since God exists necessarily, there is no time at which God fails to exist; since time exists just in case any entity exists, time has existed eternally because God has existed eternally; the same is true for space. In that sense, space and time have always existed, and are therefore uncaused.

Although space and time are uncaused, and are therefore not dependent on God's will, as bodies and minds are, Newton still takes them to depend on God's existence. There are two senses in which space and time are not necessary beings: first, they are merely contingent; and second, they are not beings at all (not even contingent ones). Space and time are contingent because they depend for their existence on the first existing being, so they do not exist as a matter of necessity unless something else exists as a matter of necessity. Of course, space and time cannot fail to exist in the sense that they exist just in case a necessarily existing entity – God – exists. But if, *per impossibile*, God and all other entities vanished, space and time would vanish as well. Hence they are not necessary beings in the sense that they depend for their existence on another being external to themselves. This partially characterizes their difference from God, who does not depend on any other being in that way.

The second point is that in addition to being contingent, space and time are also not beings, and are therefore not necessary beings. Unlike contingent beings such as ordinary objects and human minds, which (as we have seen) are substances that bear characteristic actions, space and time are causally inert and therefore fail a crucial Newtonian criterion of substancehood. Space and time are mere affections, and not genuine substances. Thus for Newton, space and time are merely contingent and are mere affections.

Carriero ends his paper with a powerful objection to Newton's view that is relevant precisely at this stage: in contending that space and time are affections of every kind of being, and that no entity can exist that is not related to space in some way, is Newton inadvertently implying that *God* depends on space and time?[30] Is it the case that God could not exist independently of space and time, and if so, does that not render space and time conditions on God's existence? As Carriero notes, Leibniz raised precisely this objection in his fifth, and last, letter to Clarke:

> If the reality of space and time is necessary for the immensity and eternity of God; if God must be in space; if being in space is a property of God [*si être dans l'espace est une propriété de Dieu*]; God will in some measure depend on time and space, and need them. (L5: 50; *Die philosophischen Schriften*, vol. VII: 403)

Although Leibniz mentions Clarke's view that space is a property of God[31] – which I (and Carriero) think Newton rejects in *De Gravitatione* – he raises a pressing objection independently of that view, and it is one that Newton must face even on my interpretation.

It should be evident from the above discussion that Newton does not think of space and time as beings – as substances – and in that sense he cannot be read as contending that God is somehow dependent upon other beings or substances. If God depends in any way on space and time – as Carriero and Leibniz allege and as I explore below – it must be agreed that God does not depend on any other substance.

Nonetheless, Carriero may still be correct in concluding that Newton characterizes God as somehow dependent on space and time, which may be problematic even if the latter are not substances. In a way, Carriero's objection presses us to clarify the affection thesis and its implications. From my point of view we might read *De Gravitatione* as providing an analysis of what it means to exist.[32] Newton seems to think that for something to exist just is for that thing to occupy space and time.[33] So if we ask whether God could exist independently of space and time, we are asking a nonsense question. To conceive of God as existing just is to conceive of God as occupying space and time. And in that sense, space and time are not

---

[30] See also Franks, *All or Nothing*, 28.

[31] As Carriero points out, Clarke tells Leibniz that space is a property of God – perhaps in a sense requiring explication – in at least two passages: C 3: 3 and C 4: 10 (*Die philosophischen Schriften*, vol. VII: 368 and 383). Carriero argues, rightly in my view, that Clarke misrepresents Newton's conception with this claim.

[32] Thanks to Michael Della Rocca for helping me to clarify this point.

[33] This might be implied by Newton's note to Des Maizeaux (quoted by McGuire, "Predicates of Pure Existence: Newton on God's Space and Time," 101), where he indicates that space and time are not qualities or properties of entities, but rather are special kinds of "modes of existence" such that if these modes are rejected, the existence of the entity in question is rejected. But my reading of this passage, and indeed of Newton's conception of space and time more generally, differs from McGuire's. He denies that Newton provides us with an analysis of existence along the lines that I have specified; for him, Newton denies that "to be is to be extended and to endure" – 103. From my point of view, however, this is precisely what Newton advances in this text.

conditions on the existence of anything; rather, they express what it means to exist. For this reason, we cannot conceive of the situation in which God exists but space and time do not, for that violates a conceptual aspect of what it means to exist.[34]

This solution potentially raises a new problem – in contending that his analysis of existence applies univocally to God and to contingent beings, Newton embraces the controversial view that we can fundamentally understand God's existence. According to one traditional position on God's existence, we can employ only analogical predication when dealing with God, even in the case of attempting to characterize God's existence. We can say only that our existence is analogous to God's existence. Among the moderns one also finds the position that God's existence is radically distinct from the existence of any other entity, so we cannot achieve even analogical predication. For defenders of either position, Newton's analysis would be controversial.

A caveat is needed here.[35] In saying that Newton provides a univocal analysis of existence in *De Gravitatione*, we should not understand him as providing a univocal analysis of being (cf. the discussion in ch. 2). For Newton does not analyze being *qua* being, and therefore does not attempt to cover all of the types that fall under the category of being – this would presumably include substances, properties, affections (e.g. space), relations, and so on. Rather, he analyzes, and provides a univocal understanding of, every kind of being, where he lists the kinds of beings as follows: material objects, the human mind, and God. This does not concern, for instance, the various properties and relations that these beings might bear, which require a separate analysis. This is sensible on Newton's part, for there is no reason to liken the existence of properties to that of things – a property such as redness presumably does not exist just in case it occupies some place at some time. On the contrary, for redness to exist, I take it that a further condition must be met, namely that there must be some thing that is red. If the analysis also held for properties then redness would presumably become akin to a substance.

The very same point can be made, *mutatis mutandis*, for affections such as space. Here we see the crucial import of Newton's denial that space and

---

[34] On this reading, Newton's understanding of existence would hold for God as well as for all other entities – the difference is that God's infinity and omniscience require that God be present at every point in space at *every* time, which ensures God's uniqueness. But the overarching view here is that we can provide a purely general analysis of existence, one that applies to God as well as every other type of entity.

[35] Thanks to Jeff McDonough for discussion of this point.

time are substances. He does not claim that space exists just in case it occupies some place – or every place? – at some time, for space is not one of the beings listed in Newton's analysis. Rather, according to Newton, for space to exist is just for it to be occupied by some kind of being. So Newton's analyses are mutually supporting: the reason that we posit space if we posit any being is that space exists just in case it is occupied by some being at some time, and any being exists just in case it occupies some place at some time. But space and entities require distinct analyses.

This still leaves us with the puzzle mentioned at the end of the first section. If my interpretation of *De Gravitatione* is correct, Newton takes space to be an affection and therefore to depend for its existence on other things. But in the Scholium to the *Principia*, as we have seen above, Newton apparently takes space to exist independently of all other things. And again, in the Scholium, Newton treats space as absolute and yet, as we have seen, in *De Gravitatione* he denies that space is "absolute in itself." How can these views be rendered consistent, if at all? This question will press us to consider not only the relation between these two texts, but more generally the relation between physics and metaphysics in Newton's eyes.

NEWTON'S ABSOLUTISM REVISITED

One way of approaching the apparent tension between *De Gravitatione* and the Scholium to the *Principia* is to reconsider the context each establishes for the discussion of space. When Newton considers the idea that space is "absolute" in the Scholium and in *De Gravitatione*, he presumably has distinct issues in mind. Very roughly put, in *De Gravitatione*, the ontology of space is at issue – hence in that text, Newton considers the idea that space is a substance, or an accident inhering in a substance, rejecting each possibility. In the Scholium to the *Principia*, however, ontology is not at issue; the focus is on the proper understanding of the distinction between absolute and relative space, and on the need to conceive of space in a way that allows us to understand absolute motion.

This suggests, in turn, that the question of whether space is absolute has distinct meanings in the two texts. In *De Gravitatione* to contemplate whether space is absolute is to contemplate the ontology of space by considering, in particular, the possibility that space is a substance. As we have seen, the relevant passage reads as follows:

Perhaps now it may be expected that I should define extension as substance, or accident, or else nothing at all. But by no means, for it has its own manner of existing which is proper to it and which fits neither substances nor accidents. It is not substance: on the one hand, because it is not absolute in itself, but is as it were an emanative effect of God and an affection of every kind of being [*non absolute per se, sed tanquam Dei effectus emanativus, et omnis enties affectio quaedam subsistit*]; on the other hand, because it is not among the proper affections that denote substance, namely actions, such as thoughts in the mind and motions in body. (*De Gravitatione*, 21)

Space is not absolute because it is an emanative effect of God. What does this contention mean? As we have seen above, to be an emanative effect of God – which follows, for Newton, from the fact that space is an affection of every kind of being and that God is the first existing being – is not to be created (or efficiently caused) by God, but to depend on God for existence. Therefore to be absolute in this context is to exist independently of every-thing, including God. This clarifies, in turn, Newton's claim that space is not a substance because it is not absolute. Newton has in mind the "Cartesian" idea that substances exist independently of whatever else exists. As we have seen, space is not a substance in this sense.[36]

But in the Scholium, the ontology of space is not at issue: we do not broach the question of whether space is a substance, so we do not question whether it is absolute in the sense just enumerated. God's relation to space – which must be considered if we are to determine whether space is depend-ent on God – is not at issue here. In the Scholium, to be absolute is to exist independently of objects and their relations. We cannot force the Scholium to answer a metaphysical question – concerning space's potential depend-ence on God – that is simply not broached there.[37]

This interpretation requires an important caveat. If one were to consider the coherence of the view in the Scholium and the affection thesis *per se*, leaving aside the other aspects of Newton's conception of space in *De Gravitatione*, one might encounter a conflict. Since the affection thesis indicates that space is an affection of ordinary objects, it indicates that space does not exist independently of objects and their relations, and is therefore not absolute in the Scholium sense. But the affection thesis does not exhaust Newton's view of space. If we consider his overarching view – including the affection thesis, its entailment (the view of emanation), his

---

[36] As we have also seen, space is not a substance in the more ordinary sense of having a characteristic type of action, for it is causally inert. So it *also* does not meet the Newtonian criterion of substancehood met by objects and by the human mind.

[37] Newton does not fail to address that issue in the Scholium because of any reluctance to discuss God in the *Principia* – on the contrary, Newton discusses God extensively in the General Scholium, and even briefly addresses God's spatial ubiquity in that section. So the point is that God's relation to space is not relevant in the Scholium, and is therefore irrelevant to the discussion of absolute space in the Scholium.

view of God, and his understanding of body – it does not conflict with the view of absolute space in the Scholium. Newton holds the familiar view that ordinary objects, including human beings, are contingent and created beings, unlike space, so before God created the universe, space existed and God occupied all of it. Were God to destroy all matter in the future, space would continue to exist unchanged. Thus the overarching view of space in *De Gravitatione* is perfectly consistent with the understanding of absolute space in the Scholium. My discussion here mirrors Newton's: he contends that space is not absolute because it is an emanative effect of God, so Newton himself takes his overarching view of space, and not simply the affection thesis, to be at issue.

There is another way of approaching this issue, one that uncovers the common ground between the discussions of space in the two texts and that can therefore be employed to characterize Newton's overarching conception of space, taking both texts into account. Consider two types of absolutism: (1) strong absolutism is the view that space exists independently of every entity; and (2) weak absolutism is the view that space exists independently of all material objects and all possible relations, but depends on God for its existence.[38] For the weak absolutist a divinely inhabited but otherwise empty universe would be spatial; but an utterly empty universe would not. For the strong absolutist even an utterly empty universe would be spatial. Now on the typical interpretation of the Scholium, of course, Newton is taken to defend strong absolutism, but the contrast with weak absolutism helps to indicate that this interpretation overreaches, attributing to the discussion in the Scholium a view of the relation between space and God. Since the Scholium is neutral on that relation, it does not express strong absolutism – to do so, it would have to assert space's independence from everything, including God. When Newton addresses the issue of God's relation to space in *De Gravitatione*, the view he defends is inconsistent with strong absolutism, since he asserts space's dependence on God in that text.[39] Thus Newton's complete view is that space is an infinite,

---

[38] Daniel Garber pointed out to me that this distinction itself mirrors Descartes's contention that "substance" cannot be applied univocally to God and to contingent beings. Descartes argues that only God is truly independent of everything else that exists; contingent beings, such as the mind, are substances in the sense that they exist independently of everything except God – see *Principia Philosophiae*, VIII-1: 24.

[39] This reading is not undermined by Newton's contention in *De Gravitatione* that we can imagine spaces that are devoid of all body, for Newton takes that to indicate only that space cannot be an accident inhering in some subject (22). Newton highlights our imaginative capacity here for a clear reason: when assessing whether something is an accident of a subject, it is sensible to ask whether we can imagine that thing to exist independently of that subject. With the redness of a book, for instance,

immobile, homogeneous, Euclidean magnitude that is independent of all material objects and relations, but dependent on God.

Weak absolutism is perfectly consistent with Newton's aims in the Scholium, and with his rejection of the Cartesian view of motion. In thinking that space depends on God for its existence, Newton does not undermine the sense in which space is absolute in the Scholium – we can still conceive of true motion as change of absolute place, even in the case of a lonely object. For according to weak absolutism, if God exists, an infinite Euclidean magnitude that is independent of all materials objects and their relations also exists – that magnitude can therefore allow us to understand true motion not as a change in object relations *à la* Descartes, but rather as a change of absolute place. Space's dependence on God does nothing to undermine this view.

Notice that Newton's laws, along with their corollaries, do not decide between weak and strong absolutism, for the laws are equally consistent with each and require neither. If we restrict ourselves only to the measurable quantities that Newton highlights, such as acceleration – including, of course, true rotation – and leave aside true velocity, which cannot be measured, we require only inertial frames; we do not require absolute space in either its weak or strong guises. Similarly Newton's laws, and especially his fifth corollary to the laws, indicate that absolute space – in both the weak and strong sense – gives rise to a quantity (absolute velocity) that cannot be measured.[40] In that sense the distinction between weak and strong absolutism concerns an issue on which Newton's physical theory would appear to be neutral. But it is crucial to understanding God's relation to space, which is central, in turn, to what I have called Newton's divine metaphysics (see ch. 6).

---

we presumably cannot imagine it to exist independently of the book; hence it is said to "inhere" in the book. But since we can imagine space to exist without any body at all, we can conclude that it is not an accident. This does not entail that space can exist independently of any entity, for Newton addresses here the question of whether space is an accident of a body, leaving aside the question of whether space can *exist* independently of all beings, which he addresses elsewhere.

[40] DiSalle puts the point elegantly, writing of absolute space:

But there it was clear even to Newton that, however absolute space may be bound up with his general metaphysical picture, it could be completely disregarded in our conception of the physical entities and processes at work in a Newtonian world: the ontology of bodies moving under the influence of accelerative forces does not require any distinction between uniform motion and rest. So the elimination of the distinction, and of absolute space, in no way disturbed the fundamental concepts of Newton's theory. (*Understanding Space–Time*, 104)

I would simply add that even if we restrict ourselves to that "metaphysical picture," Newton defends only weakly absolute space.

Just as importantly, weak absolutism is consistent with Newton's famous discussion of God and space in the General Scholium to the *Principia*:

He is not eternity and infinity, but eternal and infinite; he is not duration and space, but he endures and is present. He endures always and is present everywhere, and by existing always and everywhere he constitutes duration and space [*Durat simper, & adest ubique, & existendo semper & ubique, durationem & spatium constituit*]. Since each and every particle of space is *always*, and each and every indivisible moment of duration is *everywhere*, certainly the maker and lord of all things will not be *never* or *nowhere*. (*Principia*, 941)

Newton writes on the next page: "It is agreed that the supreme God necessarily exists, and by the same necessity he is *always* and *everywhere*" (*Principia*, 942). So although Newton does not address the question of whether space is dependent upon God here,[41] he echoes the claim in *De Gravitatione* that God is spatially ubiquitous. And his view here is consistent with weak absolutism.

This reading might be thought to conflict with Newton's contention in *De Gravitatione* that we can conceive of space as existing while feigning the non-existence of God.[42] It seems to me that this can be read as follows: for Newton, space depends upon God for its existence, for the reasons outlined above, but space and God are notionally distinct.[43] That is, we can conceive of space without conceiving of God, and we can conceive of space as existing while feigning the non-existence of the necessarily existing being.[44] This underscores once again the distinction between an affection and a property (or perhaps, a certain type of property). Perhaps some of God's properties are not notionally distinct from God – for instance, omnipotence and God might not be distinct. Perhaps to think of God just is to think of an omnipotent being, and to think of omnipotence just is to think of God. But to think of space is to think of an infinite Euclidean magnitude, and not

---

[41] The claim here that "by existing always and everywhere," God "constitutes duration and space" might be understood as expressing something akin to the view that space is an affection of every being and therefore emanates from the first existing being, viz. God (who, as Newton says in the passage above, exists necessarily). The text appears to be too sparse to decide the matter; cf. also Carriero, "Newton on Space and Time," 131 n. 23.

[42] On this point, see Stein, "Newton's Metaphysics," 271.

[43] This might be equivalent to Descartes's notion of a distinction of reason – thanks to Sean Greenberg for making this point.

[44] This indicates that from Newton's point of view, God's necessary existence is not a matter of logic – that is, it is not logically impossible for God to fail to exist, for we can, as Newton says, "feign" his non-existence. But we presumably cannot feign that a triangle has two sides. In that sense Newton might be skeptical of Descartes's revival of the ontological argument, which collapses these two cases: for Descartes in the fifth Meditation, we can no more conceive of God as non-existent than we can conceive of a triangle that lacks three sides – each is logically impossible.

to think of God, or of God's properties; one can also conceive of space without conceiving of its relation to God. If you like, it is one thing to conceive of an infinite Euclidean magnitude, and another to ask what the ontology of such a magnitude is – the latter question will ultimately bring the recognition that space is not absolute because it is dependent on God.

This interpretation, finally, indicates that the contemporary term for absolutism – "substantivalism" – can easily be misleading when applied to Newton, especially if it is to bear a meaning that is recognizable to him.[45] Since space is weakly absolute, it may be a substance only in the sense that it is a subject of predication that is not itself a predicate of anything else. And this is probably undeserving of the "substantivalist" title. Newton would certainly have denied that if space was a subject of predication but not a predicate itself, it must also be a substance in what I have been calling the "Cartesian" sense – that is, an entity that stands alone from whatever else exists. And that, of course, is the typical view called substantivalism. Certainly in the modern debate, the substantivalist does not contemplate the idea that space depends on God.

IS NEWTON'S VIEW OF SPACE METAPHYSICAL?

It should be clear from what we have seen thus far that *De Gravitatione* was intended to address questions about the ontology of space found in the writings of many mid- to late-seventeenth-century figures, including Descartes, Gassendi, More, and Barrow. One might therefore think that it is relatively uncontroversial to conclude that Newton was willing to explore the status of space within what we might call a metaphysical context. However, this perspective on Newton's treatment of space in *De Gravitatione* remains controversial. Consistent with the radical empiricist interpretation outlined in ch. 2, for instance, Howard Stein has argued that although the affection thesis addresses a question about the ontology of space, it is in fact a merely empirical claim regarding the status of space. Stein's interpretation is provocative, pressing us to consider precisely what it means to ask whether Newton defends a metaphysical position on space.[46]

---

[45] If the substantivalist thinks that space–time points exist, it is unclear that Newton would embrace that view – see DiSalle, *Understanding Space–Time*, 37–8. Apparently, the term "substantivalism" was coined in the 1920s (thanks to Nick Huggett for this point), so Newton himself would obviously have been unfamiliar with it.

[46] Although he differs from Stein on other points McGuire might endorse this interpretation. For instance, in the course of discussing Newton's view of space, including its expression in *De Gravitatione*, McGuire (*Tradition and Innovation*, 25) writes that for Newton, "all knowledge has sensory experience as its ultimate source."

As we will see, the discussion will also broach the related question of whether his position is theological.

From Stein's point of view, we ought to understand the affection thesis as a claim about space that is founded upon our perceptual experience. Discussing the relation between the affection thesis and Newton's view that space "emanates" from the first existing being, Stein writes:

> For our second question – what reason Newton thought there was that justified this view of space as an "emanative effect" of whatever exists – it is to be noted that he describes the proposition as *inferred from* a preceding one: that "no being exists or can exist that does not have relation in some way to space"; and this in turn he founds upon an enumeration of all the kinds of "beings" he takes actually to exist, and their several relations to space. In the light of this, and of the fact that there is no suggestion – here or indeed anywhere the present writer knows of in Newton's writings – of an *a priori* epistemological ground for any item of knowledge, it appears reasonable to conclude that the reason in question is an empirical one: our experience affords no grounds for a conception of real existents – beings capable of acting – that do not have an appropriate relation to space. ("Newton's Metaphysics," 269)

Stein may be right that our perceptual experience provides us with information only of spatial entities, and this may suggest that all entities are spatial, or more precisely, that all of them bear "an appropriate relation to space."[47] Even if one grants this point, however, Stein's view may conflict with an aspect of Newton's.

As should be evident from the first section above, Newton takes all of our perceptual experience with objects to be experience of relative spaces, for we obviously lack any perception of absolute space, and when we perceive a set of objects, however close or distant they may be, we *ipso facto* perceive them as inhabiting a relative space whose parameters are defined by those objects. But notice that the affection thesis holds that absolute space, rather than any relative space, is an affection of every kind of being.[48] Newton cannot intend the affection thesis to hold for relative spaces, for the latter are by definition dependent upon particular objects for their existence, which is precisely what is denied in the affection thesis. Consider one of Newton's

---

[47] As Stein is aware, Newton includes minds in his enumeration of the kinds of being that bear a relation to space, and he defends the controversial, and decidedly anti-Cartesian, view that even the human mind is essentially spatial. (The mind could not be non-extended, on pain of failing to exist, as Newton makes clear in *De Gravitatione* – I discuss this point above.) It is difficult to see how this view could be said to be grounded upon our ordinary perceptual experience, but I will leave this issue aside for the sake of argument.

[48] Throughout my discussion of Stein's view, "absolute space" should always be read as referring to weakly absolute space.

examples from the Scholium: the "space of our air" is an instance of a relative space, one whose parameters involve the surface of the earth and (perhaps) the clouds – these objects set the parameters of the relative space. Hence the space of our air is dependent upon particular objects, such as the earth; it might therefore be a property of these objects, but could not be an affection of them, since an affection is not dependent upon particular objects for its existence.

So the difficulty is that our perceptual experience seems to acquaint us only with relative spaces, however extensive they might be, and yet the affection thesis concerns the ontological status not of a given relative space, or of relative spaces *per se*, but rather of absolute space itself. It would seem that our perceptual experience is limited to informing us that space is a property of what exists (and of what we perceive), rather than an affection of what exists.

As he notes in the above quotation, Stein buttresses his interpretation of the affection thesis by contending that Newton never mentions any "*a priori* epistemological ground for any item of knowledge." Newton's extensive reading of Descartes's *Principia Philosophiae* and *Meditations*, of course, would certainly have familiarized him with a classical rationalist conception of such an epistemological ground. However, this is far from decisive, for two primary reasons. First, despite his extensive attempts at undermining Cartesian views – of space, motion, etc. – Newton never addresses overarching epistemic issues concerning the basis of our knowledge of space, either in the Scholium or in *De Gravitatione*. His silence would appear to be neutral on the issue that Stein raises. Second, Newton also never addresses the question of whether his affection thesis, or his more complete conception of space in *De Gravitatione*, is an empirical or *a priori* claim; he simply ignores this question. So this does not tell for or against Stein's view.

Stein then addresses skepticism toward his claim that Newton's conception of space in *De Gravitatione*, including his view that God is spatially ubiquitous, is empirical in character:

It might well be asked how *experience* could be said to ground Newton's assertion that "God is everywhere." But first – although the claim that God is everywhere *present in space* was a controversial one, and even somewhat dangerous to advocate – Newton thought the doctrine of the *ubiquity* or *omnipresence* of God amply founded in the tradition of revealed truth; and second, he clearly thought experience shows that *minds* can act only *where they are*; so the doctrine of God's omnipotence (likewise founded in revelation) itself entails his omnipresence. (Stein, "Newton's Metaphysics," 270)[49]

---

[49] Cf. also in this context a passage from Locke's *Essay* that Stein cites: *Essay*, 2. 23. 20.

According to Stein, Newton took experience to support the view that minds can act only where they are, and so he could employ this view to underwrite an inference from God's omnipotence – which would presumably not be in dispute among Newton's readers – to God's spatiotemporal omnipresence in a strong sense. In addition to there being no suggestion of such an inference within Newton's published and unpublished writings, however, Stein's suggestion presumably runs foul of the distinction between a finite human mind and an infinite God. Surely we cannot argue that God's power must be expressed through a spatiotemporal location because we find that the power of human minds to interact casually with bodies is so limited. For as Newton makes perfectly clear, in a passage from the General Scholium that Stein quotes, God has no body in any sense, and therefore lacks the connection to a particular material object that is characteristic of the human mind. And as far as this particular argument is concerned, it might very well be precisely that connection, in the case of the human being, that limits the action of human minds to particular locations in space.

From Stein's point of view, Newton takes "revelation," and the tradition of "revealed truth," to represent empirical sources of our knowledge of God, so this explains, in a broad sense, how Newton could take his view of divine spatial ubiquity to be supported by experience.[50] This interpretation, however, raises an important further question. Does Newton in fact think that "revelation," in the form (say) of texts such as the Bible, could serve as the

---

[50] Earlier in "Newton's Metaphysics," Stein cites a passage from the General Scholium, a section Newton added to the second (1713) edition of the *Principia*, to support his view that Newton understood even his theological ideas to be derived from experience. The passage reads in full:

As a blind man has no idea of colors, so we have no idea of the ways in which the most wise God senses and understands all things. He totally lacks any body and corporeal shape, and so he cannot be seen or heard or touched, nor ought he to be worshipped in the form of something corporeal. We have ideas of his attributes, but we certainly do not know what is the substance of any thing. We see only the shapes and colors of bodies, we hear only their sounds, we touch only their external surfaces, we smell only their odors, and we taste their flavors. But there is no direct sense and there are no indirect reflected actions by which we know innermost substances; much less do we have an idea of the substance of God. We know him only by his properties and attributes and by the wisest and best construction of things and their final causes, and we admire him because of his perfections; but we venerate and worship him because of his dominion. (*Principia*, 942)

Stein quotes part of the last sentence in the passage. However, this is far from decisive, for the paragraph of which this passage is a part begins with the claim: "It is agreed that the supreme God necessarily exists," a view, as we have seen, that also plays a crucial role in *De Gravitatione*. And it is far from clear that this can be understood as an empirical claim about God (but cf. McGuire, *Tradition and Innovation*, 34–5). See the discussion of this issue in ch. 2 above.

source of our knowledge of God's relation to space, as Newton understands that relation? There is reason to be doubtful. An important clue is found in the Scholium to the *Principia*, where Newton clarifies an aspect of his distinction between relative and true space:

Relative quantities, therefore, are not the actual quantities whose names they bear but are those sensible measures of them (whether true or erroneous) that are commonly used instead of the quantities being measured. But if the meanings of words are to be defined by usage, then it is these sensible measures which should properly be understood by the terms "time", "space", "place", and "motion", and the manner of expression will be out of the ordinary and purely mathematical if the quantities being measured are understood here. Accordingly those who there interpret these words as referring to the quantities being measured do violence to the Scriptures. And they no less corrupt mathematics and philosophy who confuse true quantities with their relations and common measures. (*Principia*, 413–14)

Recall two aspects of Newton's view in the Scholium: (1) space, time, and motion are "quantities"; and (2) relative spaces, times, and motions are "measures" of these quantities. Hence Newton writes that a relative space is "any movable measure or dimension of this absolute space," adding that "such a measure or dimension is determined by our senses."

A typical method of distinguishing between a religious, or a theological, understanding of the divine and the conception available to the natural philosopher is to say that the former analyzes God from sacred texts, and the latter from an investigation of the "phenomena." Newton himself, of course, endorses this view, referring to it implicitly in the General Scholium (*Principia*, 943). But he adds an important subtlety to it: whereas the interpretation of scripture requires the common understanding of space, time, and motion, understanding phenomena ultimately requires a mathematical understanding. And it is precisely that mathematical conception that allows us to understand a crucial aspect of God. From Newton's point of view, scripture is written in the language of the "common person," and so in interpreting any of its descriptions of space or motion, we ought to understand these as claims concerning relative spaces. To understand these claims and descriptions as holding of absolute space, according to Newton, does "violence" to the scriptures, presumably because it undermines their veracity. Thus if scripture tells us that on a certain day the sun miraculously stopped moving, we should not understand this as a change in what Newton would call the "true" motion of the sun, which would obviously have to be accompanied by various effects, but rather as a change in its relative or apparent motion, which need not be accompanied by such effects.[51]

Newton evidences an even more subtle understanding of scriptural accounts in his pre-*Principia* correspondence. In 1681 Thomas Burnet sent Newton a copy of his *Telluris Theoria Sacra*, asking his opinion of its arguments. Newton took issue with Burnet's attitude toward the truth of scriptural descriptions of nature, responding in part as follows:

As to Moses I do not think his description of the creation either philosophical or feigned, but that he described realities in a language artificially adapted to the sense of the vulgar. Thus where he speaks of two great lights I suppose he means their apparent, not real greatness. So when he tells us God placed those lights in the firmament, he speaks I suppose of their apparent not of their real place, his business being not to correct the vulgar notions in matters philosophical but to adapt a description of the creation as handsomely as he could to the sense and capacity of the vulgar.[52]

To understand this passage, it is crucial to recognize that the distinction between the mathematical, absolute perspective and the common, apparent perspective should not be collapsed into the distinction between the true and the false. A Mosaic description of the world is not a false description. On the contrary, Newton's point is that if we understand Moses to be referring to apparent space, time, and motion, then the truth of his descriptions can be rescued. Indeed, the Hebrew Bible and sacred texts can in fact be literally true if interpreted in this way, just as the statement, "The sun set last night in Jerusalem at 10 p.m." can be literally true if it is understood as a statement about apparent space, time, and motion. Certainly that comment is not a metaphor. Thus sacred texts are neither false nor only metaphorically true – as Newton tells Burnet, they are neither "philosophical" nor "feigned" – but are in fact literally true, if understood in the right way.

These aspects of Newton's views cast doubt on Stein's interpretation. Scripture could not inform us of God's ubiquity in space, since that concerns God's presence in what Newton would call space itself, rather than in any relative space. The reason is that relative spaces are understood to be finite, so they cannot accommodate God's infinity. Hence if we interpret scripture

---

[51] As John Young pointed out to me, it remains unclear how Newton would choose to interpret this scriptural description so as to render it true. Perhaps he could do so by taking the observers on the earth to move in such a way as to render the sun *apparently* motionless. In any case, Newton provides us with a principled way of approaching such delicate issues.

[52] Newton to Burnet, January 1681, *Correspondence*, vol. II: 331; cf. also 333. On Burnet's work, see Dobbs, *The Janus Faces of Genius*, 76–7, and especially Mandelbrote, "Isaac Newton and Thomas Burnet."

as referring only to relative spaces, times, and motions, we presumably cannot rely on it as a source for our understanding of the relation between God and space itself. And as we have seen, Newton's affection thesis concerns the relation between absolute space and "every kind of being," and therefore would appear to transcend the knowledge available to us in scripture.

These points, in turn, significantly complicate the oft-repeated contention that Newton defends – or perhaps arrives at – his conception of God's relation to space on theological grounds. For in order to account for scriptural claims, we require only the common – *vulgare* – conception of space, time, and motion, and in fact the mathematical conceptions of these quantities only complicate matters, for sacred texts refer to the common measures of the quantities, and not to the quantities themselves. And so we might distinguish, very roughly, between two distinct aspects of our study of the divine: first, the interpretation of scripture, which requires only the common understanding of space, time, and motion; and second, the understanding of God within natural philosophy, which requires the true understanding of space, time, and motion.[53] If theology involves an interpretation of sacred texts, and a response to some historical tradition of interpreting such texts, as was commonly thought in Newton's day, then his conception of God's relation to space is not in fact a theological one.[54]

Absolute space is in fact central to Newton's divine metaphysics. As we have seen, he consistently contends that the infinity of the divine being ought to be construed in terms of God's presence within, or occupation of, space; presumably only an infinite space can accommodate God, on this conception. This coheres, in turn, with his view that the existence of a substance ought be construed in terms of that entity's occupation of space over time. In virtue of being finite, a substance must occupy some finite space, and in virtue of being contingent, it must occupy that space for some finite period of time. In virtue of being infinite, God must occupy all space,

---

[53] Hence the discussion of the divine is bifurcated into the interpretation of scripture, on the one hand, and natural philosophy, on the other. As for the latter, Newton explicitly contends that natural philosophy concerns the causes of phenomena, and that "to discourse" about the first cause is appropriate within it. Indeed, Newton argues in the first edition – in an often-ignored passage that was struck from the second edition – that the "system of the world," the structure of the solar system, required a divine intervention because it could not have arisen from interactions among material objects in accordance with the laws of motion alone (Cohen, "Isaac Newton's *Principia*, the Scriptures, and the Divine Providence," 530). It seems evident that Newton does not take his conception of God's relation to space and time, or his conception of God's creation of the solar system, to represent an interpretation of sacred texts.

[54] In Newton's case, this is a non-trivial distinction, for there are numerous extant manuscripts indicating his extremely extensive engagement with sacred texts and with the proper canons governing their interpretation. I am grateful to Scott Mandelbrote for a discussion of these issues.

and in virtue of being necessary, God must occupy that space at all times. Thus Newton's understanding of existence and his controversial conception of God require weak absolutism for his divine metaphysics to be coherent.

In contending that Newton's affection thesis, along with his conception of God and of God's relation to space, ought to be understood as empirical views, Stein supports his more general – and very plausible – contention that Newton fundamentally rejects a Cartesian understanding of the metaphysical foundations required for physical science, following instead the quasi-Aristotelian view that metaphysics ought to "come after" physics. In Stein's eyes, Newton thinks that without the guide of secure results in physical theory, our metaphysical work would consist solely of speculation and hypothesizing. Although Stein does not argue that the affection thesis, and the overarching conception of space in *De Gravitatione*, directly reflect any results or implications of Newton's physical theory in the *Principia*, his contention that Newton's conception of space in that text is empirical suggests that the discussion in *De Gravitatione* is consistent with the spirit, if not the letter, of Newton's overarching understanding of the relation between physics and metaphysics.

Yet as we have seen, Newton presumably would not take the affection thesis, and his understanding of God's spatial ubiquity, to be empirical claims. And he does not raise the question of space's ontology in the *Principia*, nor is that question of any clear relevance to Newton's treatment of space, time, and motion in that text. So Newton's treatment of space appears to be metaphysical precisely in the sense that it is separate from his treatment of space in physics, and is not limited to what we can know about the physical world through perceptual experience, or even through empirical science. Indeed, by Newton's own lights, if we let the discussion of relative space – and of the proper interpretation of scripture – in the Scholium to the *Principia* guide our understanding of *De Gravitatione*, the latter cannot be construed as providing us with an empirical view of space.

The claim that Newton's view of space is metaphysical can now be put more precisely. His complete view of space can only be understood if we consider his contention that space depends on God; it is therefore not an aspect of what I have called his mundane metaphysics. Rather, his understanding of God's relation to space is an essential element in his divine metaphysics. So Newton's view that space is weakly absolute is not an element in a physical metaphysics in just the sense that it is not reflective of the facts about space that are expressed in, or needed to comprehend, his work in physics. In that regard, one of Newton's most famous views is in fact a crucial component in his overarching conception of God's relation to the world.

6

God and natural philosophy

Newton's rejection of myriad Cartesian and mechanist views was a revolutionary development in late seventeenth-century natural philosophy. He reconfigured the relationship between physics and metaphysics, transforming crucial metaphysical questions into empirical issues while granting others a basic autonomy from the development of empirical science. This left Newton's physics with a parallel autonomy from the metaphysical presuppositions insisted upon in various ways by the Cartesians and the mechanists. Newton's conception of natural philosophy was novel in its emphasis on employing mathematical, especially geometrical, techniques to understand how motions of various kinds arise from forces of various kinds. But his consistent contention that the study of the divine represents a proper part of natural philosophy was more commonplace. And this vision of divine metaphysics has a traditional cast to it: like many figures in his day, Newton understands the world as consisting of substances that can act only where and when they are present. Among all of those substances, only one is infinite and necessary; the rest are finite and contingent. Yet even Newton's use of traditional metaphysical categories is modified by his controversial insistence that all substances, even God, are spatiotemporal entities whose action is limited to their location within space and time. These form essential components of Newton's limited but deeply held *Weltanschauung*.

The basic elements in that *Weltanschauung* were articulated especially through Newton's dissatisfaction with Cartesian metaphysics. From Newton's point of view, Descartes's metaphysics has four crucial failures: first, it identifies body and extension, taking empty space to be impossible; second, it contends that there are two types of substance, *res cogitans* and *res extensa*, where only the latter essentially occupies space; third, since a *res cogitans* is not extended, its existence consists in something other than the occupation of a place at a time; and fourth, God's existence is fundamentally distinct from that of *res cogitans* and *res extensa*. The identification of these failures leads Newton to embrace a fundamentally different picture: first, body and extension are distinct, so empty (weakly absolute) space is perfectly possible; second, all substances are of the same type, for even the human mind and God are extended; third, all substances, including minds, exist just in case they occupy a place at some time;[1] and fourth, as a result, God's existence is fundamentally akin to that of any other substance. Newton is no monist in Spinoza's sense – there are many substances, including God, ordinary objects, and human minds – but in a crucial sense, there is only one type of substance, namely extended substance.[2] In the most general terms, then, Newton sees the world as follows: space and time follow from the existence of God, the only necessary being, and the created world of contingent beings follows from God's will. Within that world, material objects and minds are each extended substances, and each exists just in case it occupies some place at some time, as is the case for God as well.

This indicates, in turn, a further nuance to the question of whether Newton embraces what I have called a "physical metaphysics." Since crucial elements of Newton's metaphysics constitute a framework that is immune to revision from physical theory, I have argued throughout that Newton's metaphysics is not purely physical. But in another sense, his metaphysics *is* physical, for all substances are physical: all substances, whether finite or infinite, whether contingent or necessary, are physical in precisely the sense that they are spatiotemporal local actors.[3] My action at any instant is limited

---

[1] This entails the clearly anti-Cartesian view that two substances can be in the same place at the same time, since, e.g., God is present everywhere, even where (and when) other substances, such as ordinary material objects, are present. One finds the same conception in Henry More – see *Immortality of the Soul*, book I, ch. 2, sections 10–11. Newton kept a copy of this work in his personal library. The extent to which this view is coherent remains a matter of debate. For a fascinating discussion of the idea that two objects can in fact occupy the same place at the same time, see Sanford, "Locke, Leibniz and Wiggins on Being in the Same Place at the Same Time." Sanford raises important difficulties for Wiggins.

[2] As we saw in ch. 5 above, in contending that space is not a substance, Newton makes it clear that he regards both minds and objects as substances, for each bears a characteristic type of action necessary for substancehood (see *De Gravitatione*, 21). But all substances are extended, and therefore equally fall under a higher level category. Newton's view is again closely parallel to More's – see, e.g. *Immortality of the Soul*, book I, ch. 2, section 12. More also contends that God is spatiotemporally ubiquitous – see e.g. *An Antidote Against Atheism*, ch. 7 in appendix, 334ff. Thanks to Daniel Garber for discussion of this point.

[3] A caveat mentioned in the discussion of the "affection thesis" in ch. 5 bears repeating here: the claim is not that everything is physical, but rather that all substances are. Newton does not think that space and time are substances, and so he is certainly not committed to thinking of them as physical, i.e. as spatiotemporal local actors. They are not actors at all: Newton in fact denies that they are substances in part because they are causally inert (as I argued in ch. 5). So he takes metaphysical objects such as God and the mind to be physical, but not all aspects of our ontology are physical (even in that limited sense).

to my location at that instant, a fact that expresses the nature not only of my body, but of my mind as well.[4] Most strikingly, God's omnipotence and infinitude is construed in terms of God's action at any place in the universe at any moment of its history. This constitutes a radical rejection of the view – articulated often in the seventeenth century – that metaphysics concerns non-physical entities such as God and the soul. For Newton, metaphysics does not completely reflect physical theory, but it concerns precisely the same types of substance that physics concerns. The objects of metaphysics have become physical.

This picture suggests, in turn, that Newton's understanding of the first cause within natural philosophy does not represent a merely rhetorical view. In contending that we can increase our knowledge of the divine being by studying natural phenomena through the methods of the *Principia*, Newton is not merely parroting a common understanding of the religious import of natural philosophy in the late seventeenth century. His view that God does not exist beyond the bounds of the universe, but is instead intimately present to every object in nature throughout the history of the universe, underwrites his view that the study of God is part of natural philosophy. Since God is physical in a limited but crucial sense, no special metaphysical method of inquiry is required to know the divine being. This may capture the kernel of truth in the radical empiricist reading of Newton defended by Stein and DiSalle: Newton certainly rejects the view that our knowledge of God's place within the universe is innate – perhaps implanted by God – or obtained through derivation from first principles – such as Leibniz's principle of sufficient reason. But as we have seen throughout this book, his worldview, centered on God's place in the universe, is certainly not empirical in the way that physical theory is.

Newton's revolution in metaphysics may underwrite a common, and theologically desirable, conception, but it comes at an obvious cost: few seventeenth-century philosophers would characterize God as *physical*. That

---

[4] Locke may have held a similar view. In the *Essay*, for instance, he writes:

Every one finds in himself, that his Soul can think, will, and operate on his Body, in the place where that is; but cannot operate on a Body, or in a place, an hundred miles distant from it. No Body can imagine, that his Soul can think, or move a Body at *Oxford*, whilst he is at *London*; and cannot but know, that being united to his Body, it constantly changes place all the whole Journey, between *Oxford* and *London*, as the Coach, or Horse does, that carries him; and, I think, may be said to be truly all that while in motion: Or if that will not be allowed to afford us a clear *Idea* enough of its motion, its being separated from the Body in death, I think, will: For to consider it as going out of the Body, or leaving it, and yet to have no *Idea* of its motion, seems to me impossible. (*Essay*, 2.23.20)

verges on the heretical.[5] And yet there is no doubt that Newton embraces this view throughout his mature intellectual career, defending it both in print and in unpublished texts (such as letters). One must be clear about the limits of Newton's revisionism: God remains an infinite, omnipotent, and necessary being, and God's action is certainly not subject to the laws of nature, for God is obviously neither massive nor impenetrable.[6] But God occupies and is actively present at every point in space throughout the whole history of the universe, and God acts locally, just as any other substance does.

This places an old question about Newton's "design" argument regarding the solar system in a new light.[7] It seems reasonable to think that Newton rejects the explanatory closure of the physical when he appeals directly to divine action to explicate the placement of the planetary bodies relative to the sun.[8] A nuance might be added here: perhaps Newton does not reject the explanatory closure of the physical, but rather redefines the physical. One can appeal to divine influence when explicating physical facts – such as the spatial locations of the planetary bodies within the solar system – in part because such influence is understood in fundamentally physical terms. Hence in principle, we can reach conclusions about the first cause by studying natural phenomena. This will presumably push the objection back one step: Newton may not reject the explanatory closure of the physical in the usual fashion, but he rejects the explanatory closure of the

---

[5] Leibniz wrote to J. Bernoulli in December of 1715 that Newton "cherishes astonishing ideas about God," including the idea that God is extended – *Correspondence*, vol. VI: 261. Of course, Newton's conception may have affinities with the views of other seventeenth-century thinkers, especially Hobbes and Spinoza. Unfortunately we do not know what Newton thought of their views: he does not discuss either thinker in any depth, and did not own their works in his personal library. See Harrison, *The Library of Isaac Newton*, 161 for an obscure work, possibly by Hobbes, that Newton owned; he did not own any of Spinoza's works. Newton may have been aware, and skeptical, of Hobbesian views through his reading of Henry More – see Westfall, *Never at Rest*, 97 – and he apparently read Hobbes in his youth (*Never at Rest*, 89). Thanks to Karen Detlefsen and Eric Schliesser for raising this issue with me. For a more general discussion of Newton's conception of God, see Bloch, *La Philosophie de Newton*, 496–501, and 509 for a discussion of Newton's relation to Hobbes.

[6] Thus God is a spatiotemporal local actor just as any ordinary material object is, but still differs from such objects in numerous ways. Ordinary material objects are merely contingent, and therefore depend upon God for their existence, and they are essentially massive, which entails that they follow the three laws of motion (see ch. 4 for details). They are also essentially impenetrable, which entails that they reflect light and cannot occupy the same spaces simultaneously. God, in contrast, occupies all spatial locations at all times, even when they are occupied by ordinary material objects; and God's lack of impenetrability entails that no material object can act on God, preserving the one-way causal chain typical in more traditional conceptions. See, for instance, the discussion in the General Scholium – *Principia*, 941–2.

[7] Newton makes the argument in his correspondence with Bentley in the early 1690s (*Philosophical Writings*, 95–6, 101), and then repeats it in the General Scholium in 1713 – *Principia*, 942.

[8] See the illuminating discussion of this issue in Franks, *All or Nothing*, 27–8.

nomothetic, for God's action violates the laws of nature. To conclude that God's influence explains the structure of the solar system in the *Principia* is precisely to avoid treating the laws of motion as axioms – in this case, they are not "axiomata, sive leges motus."

This places another old issue in a new light. Was God's creation of the solar system's structure a miracle? In one sense, it clearly was: as we have just seen, that event did not follow the laws of nature – the three laws of motion and the law of universal gravitation – as Newton understands them. But in another sense that Newton himself emphasizes, it was not miraculous. In his anonymous "Account" of the Royal Society's report on the calculus priority dispute with Leibniz, as we have seen, he contrasts his own view with Leibniz's in the following way:

The one teaches that God (the God in whom we live and move and have our being) is omnipresent, but not as a soul of the world: the other that he is not the soul of the world, but INTELLIGENTIA SUPRAMUNDANA, an intelligence above the bounds of the world; whence it seems to follow that he cannot do anything within the bounds of the world, except by an incredible miracle. (*Philosophical Writings*, 125)

Thus Newton's conception of God as physical may raise various philosophical and theological problems, but from his point of view, it also solves a problem that hampers (what he takes to be) the Leibnizian view. In Newton's eyes, God's action on objects within the world is not miraculous because God occupies the very space and time that the objects in question also occupy. Since God's action is always local, it is fundamentally intelligible to us. Or more precisely, our limited understanding of God's action is given new parameters (see below).

We are now in a position to reconsider Newton's reluctance to invert the Cartesian system. Why did Newton refrain from rendering physics logically prior to all of metaphysics, including even our understanding of action at a distance and of God's relation to the world? Such a move would have allowed him to embrace a clean and consistent position on any potential metaphysical problem: either we transform the problem into an empirical question that can be answered through techniques in physics, or we remain silent. That is certainly an attractive view, variants of it received considerable attention in the eighteenth century, and of course it bears an important relation to twentieth-century empiricist currents in philosophy, including the views of Carnap and Quine. Unfortunately, Newton's own writings drastically limit our ability to address this issue, for he never explicitly considered a revision in his understanding of the metaphysical framework

I have been addressing. He never indicated, for instance, a willingness to reconsider his conception of God's place within the physical world. But there is one crucial exception here – Newton was confronted by the question of distant action, and indeed by one of his strongest and most insightful proponents. In the course of editing the long-awaited second edition of the *Principia* in 1713, Roger Cotes, an astronomy professor at Cambridge, raised a thoughtful objection to Newton's discussion of gravity in book III. Cotes's objection raises several significant issues, including the question of whether Newton was willing to consider the possibility that the planetary bodies acted at a distance on one another. Since Newton held Cotes in high esteem he took the objection seriously.[9]

On 18 March 1713, Cotes sent Newton a brief sketch of his projected editor's preface for the second edition, and then added a question concerning a "difficulty" he had encountered when reading the first corollary to proposition 5 of book III. In the second edition of the *Principia* book III begins with a short preface, followed by the famous *regulae philosophandi*, and then by several pages of "phenomena" involving the planetary bodies, including Jupiter's and Saturn's satellites. The bulk of book III then appears in the form of propositions, the fifth of which reads: "The circumjovial planets [Jupiter's satellites] gravitate toward Jupiter, the circumsaturnian planets gravitate toward Saturn, and the circumsolar planets gravitate toward the sun, and by the force of their gravity they are always drawn back from rectilinear motions and kept in curvilinear orbits" (*Principia*, 805). The first corollary to this proposition takes a further step toward inferring universal gravity[10] by appealing to the third law:

Corollary 1. Therefore, there is gravity toward all planets universally. For no one doubts that Venus, Mercury, and the rest are bodies of the same kind as Jupiter and Saturn. And since, by the third law of motion, every attraction is mutual, Jupiter will gravitate toward all its satellites, Saturn toward its satellites, and the earth will gravitate toward the moon, and the sun toward all the primary planets. (*Principia*, 806)

---

[9] For details on Cotes's role as editor of the *Principia*'s second edition, see Cohen, *Introduction to Newton's "Principia,"* 227–51. Cotes died tragically in 1716 at age thirty-four; Newton apparently said, "If Mr. Cotes had lived we might have known something" – quoted in Cohen, *Introduction*, 227.

[10] In proposition 7 in book III, just a few pages later, Newton already infers: "Gravity acts on all bodies universally and is proportional to the quantity of matter in each" (*Principia Mathematica*, vol. II: 576). In *Principia* (810), Cohen and Whitman translate "Gravitatem in corpora universa fieri" as "Gravity exists in all bodies universally" – but Newton uses *fieri* here rather than some form of *existere*, which appears in the General Scholium, as I emphasize in ch. 3 (see *Principia Mathematica*, vol. II: 764). Châtelet has: "*La gravité appartient à tous les corps*," which is to say, gravity "belongs to" all bodies (*Principes mathématiques*, vol. II: 21). Her translation remains in use today – see Blay, *Les "Principia" de Newton*, 70, 95.

So in this corollary, Newton explicitly applies the third law of motion – "to any action there is always an opposite and equal reaction" – to the attractions between pairs of planetary bodies in the solar system.

In his letter to Newton, Cotes refers to the third sentence (it is the third in Latin and in English) of the first corollary above, and then presents the following objection. As for Newton's words in that sentence, writes Cotes,

I am persuaded they are then true when the attraction may properly be so called, otherwise they may be false. You will understand my meaning by an example. Suppose two globes *A & B* placed at a distance from each other upon a table, & that whilst the globe *A* remains at rest the globe *B* is moved towards it by an invisible hand; a bystander who observes this motion but not the cause of it, will say that the globe *B* does certainly tend to the center of the globe *A*, & thereupon he may call the force of the invisible hand the centripetal force of *B* & the attraction of *A* since the effect appears the same as if it did truly proceed from a proper & real attraction of *A*. But then I think he cannot by virtue of this axiom [the third law – recall that the laws of motion are called *axiomata, sive leges motus*] conclude contrary to his sense & observation that the globe *A* does also move towards the globe *B* & will meet it at the common center of gravity of both bodies. This is what stops me in the train of reasoning by which I would make out as I said in a popular way your 7th proposition of the third book. I shall be glad to have your resolution of the difficulty, for such I take it to be. If it appears so to you also, I think it should be obviated in the last sheet of your book which is not yet printed off or by an *addendum* to be printed with the errata table. For till this objection be cleared I would not undertake to answer any one who should assert that you do *hypothesim fingere*, I think you seem tacitly to make this supposition that the attractive force resides in the central body.[11]

Cotes indicates, among other things, that we can apply the third law to the gravitational interactions between any two planetary bodies – such as Jupiter and one of its satellites – if we know that they act directly on one another. This is apparent in Cotes's claim that we can apply the third law if the attraction between the two bodies is one "properly so called."[12] If there were to be another body interacting with them – such as Cotes's "invisible hand," pushing (say)

[11] Cotes to Newton, 18 February 1713, *Correspondence of Sir Isaac Newton and Professor Cotes*, 152–3. Cotes's discussion is very precise: from Newton's point of view, gravity can be considered a centripetal force independently of any characterization of its physical "cause" since it is centripetal if it is a force "whereby a body tends toward a point as to a center." This leaves it open whether (e.g.) there is some medium between the body and the center in question. Hence in Cotes's example of the two globes, even if an "invisible hand" acting by impact on globe B pushes it toward globe A, we can still attribute a centripetal force to B (even from the perspective of an agent who can perceive the hand). See definition five at *Principia*, 405–6.

[12] Astonishingly, Cotes's description of an attraction "properly so-called" echoes the language that Leibniz himself uses a few years later when criticizing Newton. For instance, in his last letter to Clarke he writes:

the satellite toward Jupiter – then we would have to apply the third law to the interactions between the satellite and the invisible hand. In that case, the "attraction" between Jupiter and its satellite would be merely apparent, and the third law would not apply to these two bodies in the right way.[13]

Cotes's objection goes to the heart of Newton's argument for universal gravity, since his argument relies on the very application of the third law of motion that Cotes questions. How does Newton know that an "attractive force" resides in Jupiter? Can he rule out the possibility that some other entity pushes the satellite toward it? The objection raises important questions about Newton's methodology, but it can also be understood as raising the question of action at a distance. In modern terms, Cotes indicates that we can apply the third law of motion, as Newton does, to any two planetary bodies only if they exchange momentum with one another directly. This does not rule out the possibility that two spatially separated planetary bodies, such as Jupiter and Io (one of its "Galilean" satellites), exchange momentum with one another through some kind of medium. What Cotes's point requires is simply that there is no momentum "leakage" into the medium, for in that case, we would be forced to apply the third law to the system including the two bodies and the medium. And of course, as Cotes knew, Newton lacked evidence for any such medium.

Since Newton lacked evidence for a medium among the planetary bodies, and since it would seem *ad hoc* to postulate that there is such a medium and that it always preserves the momentum exchanges between all of the planetary bodies (and can therefore be ignored in our applications of the third law), Cotes's objection indicates a strong reason for Newton to address

---

For it is a strange fiction [*étrange fiction*] to make all matter gravitate, and that toward all other matter, as if all bodies equally attract all other bodies according to their masses and distances, and this by an attraction properly so called [*une attraction proprement dite*], which is not derived from an occult impulse of bodies, whereas the gravity of sensible bodies toward the center of the earth ought to be produced by the motion of some fluid. (*Die philosophischen Schriften*, vol. VII: 397–398)

Cotes informed Newton of Leibniz's letter to Nicholas Hartsoeker published in the *Memoirs of Literature* in May of 1712 – see Cotes to Newton, 18 February 1713, *Correspondence of Sir Isaac Newton and Professor Cotes*, 153. Cotes indicates in this letter that he will respond to some of Leibniz's objections – without mentioning Leibniz by name – in his editor's preface to the second edition of the *Principia*. Newton then wrote a response to Leibniz's letter, but his remarks were never published during his lifetime – see *Philosophical Writings*, 114–17.

[13] For discussions of the Cotes objection and of related issues in understanding the argument of book III leading up to proposition 7, see Koyré, *Newtonian Studies*, 273–83; Stein, "From the Phenomena of Motions to the Forces of Nature"; and, Densmore, *Newton's Principia*, 427–8. See also the illuminating discussion in Friedman, "Kant and Newton: Why Gravity is Essential to Matter," which then links Cotes's objection to Kant's views in an especially intriguing way. Koyré in particular is insightful in indicating the depth of Cotes's insight, and in recognizing Newton's failure to respond to the objection adequately.

the question of action at a distance. If he were open to considering the possibility that the planetary bodies act directly on one another across empty space, then this would represent an obvious place to make that explicit. For if he were to endorse even the possibility of such action, Newton could indicate that he was committed to the application of the third law – which is crucial to his theory of universal gravity – even if no medium between the planets were ever found. This would enable Newton to make a familiar maneuver, separating his physical theory and its application to the bodies in the solar system on the one hand from his own preferred interpretation of that theory and its implications for our understanding of action in nature on the other. It would enable him to render explicit the fact that he himself will not feign his hypothesis of a medium.[14] In the process of editing the very edition of the *Principia* in which he directly contests the mechanical philosophy, then, Newton was forced to confront the fact that his arguments might imply an even more dramatic rejection of mechanism.

Given the high philosophical stakes, Newton's response to Cotes's objection, written only days later, is bound to be disappointing. Newton begins by evading Cotes's point, clarifying what he means by calling a given claim a hypothesis. He then responds to the objection as follows:[15]

Now the mutual & mutually equal attraction of bodies is a branch of the third law of motion & how this branch is deduced from the phenomena you may see in the end of the corollaries of the laws of motion, pag. 22 [*Principia*, 430]. If a body attracts another body contiguous to it & is not mutually attracted by the other: the attracted body will drive the other before it & both will go away together with an accelerated motion in infinitum, as it were by a self moving principle, contrary to the first law of motion, whereas there is no such phenomenon in all nature.[16]

---

[14] In an obviously different historical and philosophical context, Kant raises a strikingly similar objection, but then uses it to criticize Newton's reluctance to endorse distant action – see especially remark two to proposition 7 of ch. 2 in *Metaphysische Anfangsgründe der Naturwissenschaft*, Akademie IV: 515–16. For an especially illuminating treatment of these issues, see Friedman, "Kant and Newton: Why Gravity is Essential to Matter," and the extended discussion in *Kant's Construction of Nature*.

[15] Cf. the discussion in Hesse, *Forces and Fields*, 137, and in Koyré, *Newtonian Studies*, 276.

[16] Newton to Cotes, 28 March 1713, *Correspondence of Sir Isaac Newton and Professor Cotes*, 155. Newton wrote a draft of his letter that he never sent to Cotes – in that version, he begins to address Cotes's objection directly when he writes: "What has been said, doth not hinder the body B from being moved by an invisible hand towards the resting body A" (*Philosophical Writings*, 122), but then the letter ends here, mid-sentence. In a further letter sent to Cotes on 31 March 1713, Newton mentions the issue raised by Cotes, but does not address it further. In his letter of 28 March, Newton cites the Scholium to the laws of motion where he discusses the applicability of the third law to attractions (*Principia*, 427–8), noting that a violation of the third law will result in a violation of the first law.

Newton indicates how a violation of the third law can lead to a violation of the first law, but he dodges the question of how we can justifiably apply the third law to cases in which two bodies are not contiguous, but rather spatially separated. For in that case, as Cotes indicates, we presumably must know what, if anything, lies between the bodies in question, at least in so far as it may exchange momentum with those bodies. And it seems that Newton simply evades that question. He therefore never confronts the heart of Cotes's analysis: does the application of the third law to spatially separated bodies suggest that they might act directly on one another through empty space? Is that a possibility one must consider?

If Newton were ever to contemplate a radical revision of his conception of action on the basis of empirical evidence, this would surely be the occasion. For Cotes has objected to the application of what Newton himself takes to be an empirical principle – the third law of motion – to the interactions among the planetary bodies in the solar system, as these were understood using observational data then available. There could not be a clearer case in which the course of empirical science might be thought to conflict directly with a metaphysical conception of how bodies in the world act on one another. Whereas Newton's theory seems to suggest that one must take the possibility of action at a distance seriously, Newton himself rejects such action as "inconceivable." This episode mirrors that in which Newton confronts the *a priori* contact principle of the mechanists with his empirical conclusion that the force of gravity is not mechanical because it is proportional to mass. And yet Newton evades the crucial issue: he never indicates a willingness to consider that action at a distance might be possible after all.[17] His response thereby differs fundamentally from Cotes's own, for in his original preface to the second edition of the *Principia*, Cotes argues that gravity is essential to matter, a claim, as we have seen, that Newton would have taken to entail action at a distance between material bodies. Unlike Newton, then, Cotes was certainly willing to embrace the possibility of distant action.[18]

---

[17] From the point of view of some commentators, Newton evades this issue at the cost of allowing a crucial tension to enter his thought. For instance, Kant claims that the argument regarding universal gravity in book III of the *Principia*, coupled with Newton's insistence that he has not embraced distant action and has yet to uncover the "cause" of gravity, sets Newton at variance with himself. See Kant, *Metaphysische Anfangsgründe der Naturwissenschaft*, Akademie IV: 515.

[18] For Cotes's claim that gravity is essential to matter, see *Correspondence of Cotes and Newton*, 158; also available in *Correspondence*, vol. V: 412–13. Cotes may in fact have been the first Newtonian to embrace action at a distance. Maxwell presents this interpretation, carefully noting that Newton himself never accepted this view – "On action at a distance," *Scientific Papers*, Vol. II: 315–16. As evidence, Maxwell cites Cotes's editor's preface to the second edition of the *Principia*, which I discuss in ch. 4 above.

This episode illustrates the crucial difference in Newton's eyes between the question of whether some causation is – *contra* Leibniz, Huygens *et al.* – non-mechanical and the question of whether action at a distance occurs anywhere in nature. Newton rejects the mechanist perspective with the revolutionary suggestion that the question concerning causation must be treated through empirical means, and in particular through the development of a physical theory that attributes observable motions within the solar system to the force of gravity. But he apparently never regards the question of action at a distance to be empirical in this way – despite Cotes's insightful argument, and the strong evidence involving Newton's own application of the (empirical) third law of motion to the interactions of the planetary bodies, he never accepts the possibility of distant action. Hence Newton's understanding of action, like his understanding of God's place within the physical world, forms a metaphysical framework for his thinking in precisely the sense that it is not subject to revision through reflection on experience or through the development of empirical science.

Since Newton surely understood Cotes's objection, the most reasonable conclusion is that he was not willing to consider any alteration in a fundamental metaphysical principle on the basis of evidence from physical theory. As for *why* he chose that response, we can only speculate. Given the arguments throughout this book there are at least two routes to understanding Newton's attitude. First, recall especially Newton's letter to Bentley in 1693, and his discussion of Leibniz's conception of God in the "Account" of 1715 – in these texts, Newton explicitly contends that action at a distance is "inconceivable," and that if God were distant from the physical world, divine intervention would be rendered an "incredible miracle," a phenomenon beyond belief. Since these strong remarks were made both before and after Cotes raised his objection it seems reasonable to conclude that Newton could not consider the possibility that the reason the third law applied to planetary interactions was that the planets acted directly on one another across empty space. Perhaps no amount of evidence from physical theory could lead Newton to embrace the possibility of an "inconceivable" action between material bodies that he refused to attribute even to the divine being.

Secondly and more significantly, we can also highlight what would be at stake if Newton *were* to contemplate a fundamental revision in his understanding of physical action *à la* Cotes. For as we have seen, Newton accords God a kind of metaphysical primacy in his worldview, and throughout his mature career, Newton always contends that God is spatiotemporally ubiquitous and actively omnipresent. For Newton to accept the suggestion that we can apply the third law of motion to the planetary bodies even if there is no medium between them would be for him to imagine attributing an entirely new kind of action to material bodies. That move, in turn, would pressure him to rethink his understanding of God's action within the world – if material bodies act at a distance on one another, surely God is not limited to local action. But this would presumably conflict with the discussion in his "Account"; moreover, if God acts non-locally, a principal reason for characterizing God as spatiotemporally ubiquitous is removed. If divine distant action is possible, then God's omnipotence need not be construed as Newton always construes it, in terms of divine omnipresence. And that move, finally, obviates the need for Newton to insist that space itself is infinite and exists independently of material objects, for we no longer require an infinite space to accommodate an infinite and omnipresent God. Since these elements in Newton's metaphysical framework are closely intertwined, removing one of them threatens to undermine the entire structure. Therefore, for Newton to contemplate a radical revision of his conception of action is for him to consider an entirely new understanding of God's relation to space, time, and the physical world.[19]

These reflections on God's relation to the world, in turn, suggest a further nuance in Newton's attitude toward the idea that nature involves only mechanical causation. As we have seen Newton rejects that view on fundamentally empirical grounds. But his careful wording of that rejection exhibits a subtlety that is especially relevant here.[20] In the General Scholium of 1713, Newton contends that gravity's proportionality to mass indicates that it does not act on surfaces, "as mechanical causes are wont to do." This formulation leaves open the possibility that gravity is mechanical in some other sense that Newton's readers may find significant – for instance, perhaps it is mechanical in the sense that it follows the laws of motion, or in the sense that it involves only local action. Various mechanists may insist on such points. Newton's openness to these points is also evident in his "Account" of 1715, where he (polemically) contrasts his views with Leibniz's: "The one for want of experiments to decide the question doth not affirm whether the cause of gravity be mechanical or not mechanical: the other that

---

[19] My discussion here has been influenced by Michael Friedman – see especially his paper, "Newton and Kant on Absolute Space." From Friedman's point of view, given Newton's views of space and of God, and given his rejection of action at a distance, the most natural way for him to understand the direct momentum exchange between the planetary bodies, which he requires for his argument concerning universal gravity in book III, is to postulate an immaterial medium between them, such as God himself. This is stronger than the claim I make here, but consistent with it.

[20] Many thanks to George Smith for a memorable discussion of this issue.

it is a perpetual miracle if it be not mechanical" (*Philosophical Writings*, 125). Of course, Newton is not retracting his conclusion that gravity is proportional to mass, which indicates a sense in which it, and its "cause," are not mechanical; rather, he is leaving it open whether gravity is mechanical in some other important sense. He presents this as an empirical question.

Contrast this discussion of mechanical causation within nature with Newton's discussion of the "first cause." We find immediately that Newton does not exhibit a parallel willingness to leave a characterization of God's causation open to empirical investigation. In query 28 to the *Opticks* – added to the first Latin edition of the text in 1706 and retained in all future editions – Newton rejects the Cartesian "medium" of planetary vortices, connecting his rejection to what he regards as an aspect of ancient wisdom:

And for rejecting such a medium, we have the authority of those the oldest and most celebrated philosophers of Greece and Phoenicia, who made a vacuum and atoms, and the gravity of atoms, the first principles of their philosophy; tacitly attributing gravity to some other cause than dense matter. Later philosophers banish the consideration of such a cause out of natural philosophy, feigning hypotheses for explaining all things mechanically, and referring other causes to metaphysics: whereas the main business of natural philosophy is to argue from phaenomena without feigning hypotheses, and to deduce causes from effects, till we come to the very first cause, which certainly is not mechanical. (*Opticks*, 369)

Whereas the question of whether gravity, and other causes of natural phenomena, must operate mechanically is always open to empirical investigation, it seems that the characterization of the first cause is not. For God's causal action is "certainly" not mechanical. Newton shows no willingness to reconsider the claim that God is not a mechanical cause – he also provides no indication of how empirical research might bear on this issue. The contrast with his work on gravity is stark.

Perhaps this aspect of Newton's thought should not be surprising. For in the late seventeenth century what does it mean to have a concept of God as the creator of the universe? It presumably means to have a concept of a necessary being, an infinite substance that causes phenomena in a way that is fundamentally distinct from any other causation within nature. To leave it open whether God might in fact be a contingent being, or a merely finite substance, or a merely "mechanical" cause in some significant sense, just is to leave open the possibility that God does not exist. It is to leave open the possibility that the world simply lacks any first cause in a recognizable sense. For these ideas concerning God's status as the first cause would seem to cut across numerous philosophical positions and fault lines. In Newton's intellectual context, to question whether any being is necessary, any substance is infinite, any cause is the first one (and therefore distinct from any "secondary cause"), just is to open a path to atheism. And it would be wholly unfair to suggest that Newton ever took that possibility seriously.

It is Newton's conception of God that drives his considered view of absolute space and of physical action alike. For as we have seen, Newton's physical theory and its laws of nature are perfectly compatible with the postulation of action at a distance, and with a weaker conception of space than the kind of absolutism Newton defends in the *Principia* (and even in *De Gravitatione*). As Newton himself was in a position to recognize, his physical theory provides one with no reason to reject action at a distance, and it provides no grounds for thinking that space must be absolute (in either the weak or strong sense). Yet Newton's conception of God very clearly requires a denial of action at a distance and the postulation of (weakly) absolute space. Thus it is precisely that conception that is crucial for assessing his considered views in these areas, and it is that conception that determines the place of action and of space within his divine metaphysics.

Newton conceived of his avowedly controversial conception of God's spatiotemporal ubiquity as enabling one to achieve a limited understanding of God's action within the physical world. He proposed a limited analogy between the human will's control over the body on the one hand, and the control God's will exercises over all of nature on the other. The "uniformity in the bodies of animals," he writes in query 31 to the *Opticks*, can be

the effect of nothing else than the wisdom and skill of a powerful ever-living agent, who being in all places, is more able by his will to move the bodies within his boundless uniform sensorium, and thereby to form and reform the parts of the universe, than we are by our will to move the parts of our own bodies. And yet we are not to consider the world as the body of God, or the several parts thereof, as the parts of God. He is an uniform being, void of organs, members or parts, and they are his creatures subordinate to him, and subservient to his will; and he is no more the soul of them, than the soul of man is the soul of the species of things carried through the organs of sense into the place of its sensation, where it perceives them by means of its immediate presence, without the intervention of any third thing. The organs of sense are not for enabling the soul to perceive the species of things in its sensorium, but only for conveying them thither; and God has no need of such organs, he being every where present to the things themselves. (*Opticks*, 403)[21]

---

[21] Some of the members of Newton's intellectual circle at this time, such as Samuel Clarke, defended a remarkably similar view. See, for instance, Clarke's claim that God's infinity entails that God "must be everywhere present" in his Boyle Lectures of 1704, *A Demonstration of the being and attributes of God*, 80. These lectures were presented two years before Clarke's translation of Newton's *Opticks* into Latin appeared; the latter contained the passage quoted above, which eventually became query 31 in later editions of the text.

This was added to the 1706 (Latin) edition of the *Opticks* and retained in all future editions. Seven years later, in the second edition of the *Principia*, Newton provided an equally famous description in the new General Scholium:

All the diversity of created things, each in its place and time, could only have arisen from the ideas and the will of a necessarily existing being. But God is said allegorically to see, hear, speak, laugh, love, hate, desire, give, receive, rejoice, be angry, fight, build, form, construct. For all discourse about God is derived through a certain similitude from things human, which while not perfect is nevertheless a similitude of some kind. This concludes the discussion of God, and to treat of God from phenomena is certainly a part of natural philosophy. (*Principia*, 942–3)

We cannot hope to comprehend God, or God's action, fully. But we can take one step toward comprehension by conceiving of God as being present to each part of each object throughout nature, able to move those objects in a way that is akin to the way that I move my own hand. The analogy for Newton is somewhat stricter than it would be for a Cartesian, for as we have seen, Newton takes the human mind to be an extended substance just as surely as the body is. So my mind is spread out in space, present in the very place that my body occupies. And God bears something akin to that relationship to all of nature, throughout history.

These reflections help to illuminate Newton's contention in the General Scholium that natural philosophy can legitimately include a discussion of God. As we have seen above, the natural philosopher is concerned with the actual phenomena present in nature; traditionally, the human mind, or the soul, was understood to be one such phenomenon. Newton did not disagree. So by presenting the idea that the mind is an extended substance, one characterized essentially by thinking, and by articulating an analogy between the mind's relation to the body and God's relation to the creation, Newton helps to indicate how these issues form a proper subject for natural philosophers. There is a genuine sense in which God and the mind are in nature. But as the passage from the General Scholium makes clear, the analogy, like any analogy, has its limits. So even on Newton's view, our ignorance of the divine remains deep. In particular, we are ignorant of how an infinite penetrable extended substance is able to interact causally with finite impenetrable extended substances. One source of our ignorance, of course, is that we do not understand how a finite penetrable extended substance, the mind, can interact causally with a finite impenetrable extended substance, the body.[22] But Newton seems to suggest that our

---

[22] See also Newton's discussion of the relation between the body and the will (in the case of animals) in his unpublished letter to the editor of the *Memoirs of Literature* in May of 1712, *Philosophical Writings*, 117.

ignorance of the one is somehow parallel to our ignorance of the other. He marshals his controversial views to take one step toward a fuller understanding of the divine being. Newton may have intended his views as helping to explain the idea that we are created in the image of God.

The perspective from which one views these aspects of Newton's thought will surely determine how one ultimately assesses them. If one views them from the perspective of future developments – especially from the point of view of a "Newtonian" in the eighteenth century – then they can easily seem to represent a dogmatic streak in Newton's work. Later Newtonians were willing to take precisely the steps that Newton balked at, embracing action at a distance as a perfectly intelligible means of natural change, dispensing with the concept of substance altogether, rejecting the metaphysical primacy of the divine being.[23] From their perspective, one might say, Newton was unwilling to embrace the most radical implications of his own rejection of Cartesianism and mechanism in natural philosophy, holding fast to an outmoded metaphysical picture of God's relation to – and action within – the physical world. But if one views them from within the intellectual milieu in which they developed, comparing them to the views of Newton's most prominent predecessor, Descartes, and to his most vociferous interlocutor, Leibniz, then the assessment changes substantially. For these defenders of the mechanical philosophy, despite their substantive disagreements in some arenas, articulated a fundamentally different conception of the relation between physics and metaphysics than Newton. From their perspective, Newton's challenge to their views was nothing short of astonishing. If we adopt this perspective, our conclusion might be that Newton's work was indeed revolutionary – but as with any revolution, it had its limits.

[23] See the illuminating discussion in Koyré, *Newtonian Studies*, 18–21. In some respects, Koyré's interpretation of Newton parallels my own.

Bibliography

Aiton, Eric, *The Vortex Theory of Planetary Motions*, New York: American Elsevier, 1972.

Anonymous [Isaac Newton], "An Account of the Book Entitled *Commercium Epistolicum Collinii & aliorum, De Analysi promota*," *Philosophical Transactions of the Royal Society* (January–February 1714), 173–224.

Ayers, Michael, "Mechanism, Superaddition, and the Proof of God's Existence in Locke's Essay," *The Philosophical Review* 90 (1981), 210–51.

Barrow, Isaac, *The Usefulness of Mathematical Learning Explained and Demonstrated*, London: Austen, 1734.

The Geometrical Lectures of Isaac Barrow, translated by J. M. Child, Chicago: Open Court Press, 1916.

Bentley, Richard, *Eight Boyle Lectures on Atheism, 1692*, New York: Garland, 1976.

The Correspondence of Richard Bentley, ed. Christopher Wordsworth, London: John Murray, 1842.

Berkeley, George, *Philosophical Works*, ed. M. R. Ayers, London: Dent, 1975.

Principles of Human Knowledge, ed. Roger Woodhouse, London: Penguin, 1988.

De Motu and the Analyst, ed. and trans. Douglas Jesseph, Dordrecht: Kluwer, 1992.

Bertoloni Meli, Domenico, *Equivalence and Priority: Newton vs. Leibniz*, Oxford: Oxford University Press, 1993.

Thinking with Objects: The Transformation of Mechanics in the Seventeenth Century, Baltimore: Johns Hopkins University Press, 2006.

"Inherent and Centrifugal Forces in Newton," *Archive for History of Exact Sciences* 60 (2006), 319–35.

Blackwell, Richard, "Descartes' Concept of Matter," in Ernan McMullin (ed.), *The Concept of Matter in Modern Philosophy*, Notre Dame: University of Notre Dame Press, 1978.

Blair, Ann, "Natural Philosophy," in Katharine Park and Lorraine Daston (eds.), *The Cambridge History of Science*. vol. III: *Early Modern Science*, Cambridge: Cambridge University Press, 2006.

Blay, Michel, *Les "Principia" de Newton*, Paris: Presses universitaires de France, 1995.

Bloch, Léon, *La Philosophie de Newton*, Paris: Librairies Félix Alcan, 1908.

Boas, Marie, "The Establishment of the Mechanical Philosophy," *Osiris* 10 (1952), 412–541.

Böhme, Gernot, "Philosophische Grundlagen der Newtonschen Mechanik," in K. Hutter (ed.), *Die Anfänge der Mechanik. Newtons Principia gedeutet aus ihrer Zeit und ihrer Wirkung auf die Physik*, Berlin: Springer Verlag, 1989.

Boyle, Robert, *A Free Enquiry into the Vulgarly Received Notion of Nature*, in *The Works of the Honourable Robert Boyle*, London: W. Johnston *et al.*, 1772, vol. IV.

Broackes, Justin, "Substance," *Proceedings of the Aristotelian Society* (2005–6), 131–66.

Broughton, Janet, "Hume's Ideas about Necessary Connection," *Hume Studies* 13 (1987), 217–44.

Buchdahl, Gerd, "History of Science and Criteria of Choice," in Roger Stuewer (ed.), *Minnesota Studies in the Philosophy of Science*, vol. V, Minneapolis: University of Minnesota Press, 1970.

  "Explanation and Gravity," in Mikulas Teich and Robert Young (eds.), *Changing Perspectives in the History of Science*, Dordrecht: D. Reidel, 1973.

Cantor, G. N. and M. J. S. Hodge (eds.), *Conceptions of Ether: Studies in the History of Ether Theories, 1740–1900*, Cambridge: Cambridge University Press, 1981.

Carnap, Rudolf, *Introduction to the Philosophy of Science*, New York: Basic Books, 1966.

Carriero, John, "Newton on Space and Time: Comments on J. E. McGuire," in Phillip Bricker and R. I. G. Hughes (eds.), *Philosophical Perspectives on Newtonian Science*, Cambridge, MA: MIT Press, 1990.

Casini, Paolo, "Newton: The Classical Scholia," *History of Science* 22 (1984), 1–58.

Charleton, Walter, *Physiologia Epicuro-Gassendo-Charltoniana*, London: 1654.

Clarke, Desmond, *Occult Powers and Hypotheses: Cartesian Natural Philosophy under Louis XIV*, Oxford: Oxford University Press, 1989.

Clarke, Samuel, *A Demonstration of the being and attributes of God*, ed. Ezio Vailati, Cambridge: Cambridge University Press, 1998.

Clarke, Samuel and Gottfried Wilhelm Leibniz, *A collection of papers, which passed between the late learned Mr. Leibnitz and Dr. Clarke, in the years of 1715 and 1716*, London: Printed for James Knapton, 1717.

Cohen, I. Bernard, "Roemer and the First Determination of the Velocity of Light (1676)," *Isis* 31 (1940), 327–79.

  "'Quantum in se est': Newton's Concept of Inertia in Relation to Descartes and Lucretius," *Notes and Records of the Royal Society of London* 19 (1964), 131–55.

  "Hypotheses in Newton's Philosophy," *Physis. Rivista Internazionale di Storia della Scienza* 8 (1966), 163–84.

  "Isaac Newton's *Principia*, the Scriptures, and the Divine Providence," in Sidney Morgenbesser, Patrick Suppes, and Morton White (eds.), *Philosophy, Science and Method: Essays in Honor of Ernest Nagel*, New York: St. Martin's Press, 1969.

  Introduction to Newton's "Principia," Cambridge, MA: Harvard University Press, 1971.

  The Newtonian Revolution, Cambridge, MA: Harvard University Press, 1980.

"Newton's Copy of Leibniz's *Théodicée*," *Isis* 73 (1982), 410–14.

"Newton and Descartes," in Giulia Belgioioso *et al.* (eds.), *Descartes. Il metodo e i saggi*, Rome: Istituto della Enciclopedia Italiana, 1990.

"The Review of the First Edition of Newton's *Principia* in the *Acta Eruditorum*, With Notes on the Other Reviews," in P. M. Harman and Alan Shapiro (eds.), *The Investigation of Difficult Things: Essays on Newton and the History of the Exact Sciences in Honour of D. T. Whiteside*, Cambridge: Cambridge University Press, 1992.

"Newton's Concepts of Force and Mass," in I. Bernard Cohen and George Smith (eds.), *The Cambridge Companion to Newton*, Cambridge: Cambridge University Press, 2002.

Cohen, I. Bernard and Alexandre Koyré, "The Case of the Missing *Tanquam*: Leibniz, Newton and Clarke," *Isis* 52 (1961), 555–66.

"Newton and the Leibniz–Clarke Correspondence," *Archives internationales d'histoire des sciences* 15 (1962), 63–126.

Cohen, I. Bernard and George Smith (eds.), *The Cambridge Companion to Newton*, Cambridge: Cambridge University Press, 2002.

Costabel, Pierre, "Newton's and Leibniz' Dynamics," in Robert Palter (ed.), *The Annus Mirabilis of Sir Isaac Newton, 1666–1966*, Cambridge, MA: MIT Press, 1970.

Costello, William, *The Scholastic Curriculum at Early Seventeenth-Century Cambridge*, Cambridge, MA: Harvard University Press, 1958.

Cunningham, Andrew, "How the *Principia* Got its Name; or, Taking Natural Philosophy Seriously," *History of Science* 29 (1991), 377–92.

"The Identity of Natural Philosophy: A Response to Edward Grant," *Early Science and Medicine* 5 (2000), 259–78.

Danielson, Dennis, "Scientist's Birthright," *Nature* 410 (26 April 2001), 1031.

De Gandt, François, *Force and Geometry in Newton's "Principia,"* trans. Curtis Wilson, Princeton: Princeton University Press, 1995.

"Newton's Nature: Does Newton's Science Disclose Actual Knowledge of Nature?" *St. John's Review* 45 (1999), 7–19.

De Pierris, Graciela, "Hume and Locke on Scientific Methodology: The Newtonian Legacy," *Hume Studies*, Forthcoming.

Densmore, Dana, "Cause and Hypothesis: Newton's Speculation about the Cause of Universal Gravitation," *St. John's Review* 45 (1999), 94–111.

Newton's Principia: The Central Argument, 3rd edn, Santa Fe: Green Lion Press, 2003.

Des Maizeaux, Pierre, *Recueil de diverses pieces, sur la philosophie, la religion naturelle, l'histoire, les mathematiques, &c.*, Amsterdam: Chez H. Du Sauzet, 1720.

Descartes, René, *Principia Philosophiae*, in *Œuvres de Descartes*, ed. Charles Adam and Paul Tannery, vol. VIII-1, Paris: Vrin, 1982.

The Philosophical Writings of Descartes, trans. John Cottingham, Robert Stoothoff, and Dugald Murdoch, vol. I, Cambridge: Cambridge University Press, 1985.

Principes, in *Œuvres de Descartes*, ed. and trans. Charles Adam and Paul Tannery, vol. IX-2, Paris: Vrin, 1989.

Principles of Philosophy, trans. and ed. V. R. and R. P. Miller, Dordrecht: Kluwer, 1991.

DiSalle, Robert, "Newton's Philosophical Analysis of Space and Time," in I. Bernard Cohen and George Smith (eds.), *The Cambridge Companion to Newton*, Cambridge: Cambridge University Press, 2002.

Understanding Space–Time, Cambridge: Cambridge University Press, 2006.

Dobbs, Betty Jo Teeter, "Newton's Rejection of the Mechanical Aether," in Arthur Donovan *et al.* (eds.), *Scrutinizing Science*, Dordrecht: Kluwer, 1988.

The Janus Faces of Genius: The Role of Alchemy in Newton's Thought, Cambridge: Cambridge University Press, 1991.

"Newton as Final Cause and First Mover," *Isis* 85 (1994), 633–43.

Domski, Mary, "The Constructible and the Intelligible in Newton's Philosophy of Geometry," *Philosophy of Science* 70 (2003), 1114–24.

Downing, Lisa, "Locke's Newtonianism and Lockean Newtonianism," *Perspectives on Science* 5 (1997), 285–310.

"The Status of Mechanism in Locke's *Essay*," *The Philosophical Review* 107 (1998), 381–414.

Duhem, Pierre, *Sozein ta Phainomena. Essai sur la notion de théorie physique de Platon à Galilée*. Paris: A. Hermann, 1908.

Edleston, J. (ed.), *Correspondence of Sir Isaac Newton and Professor Cotes*, London, 1850.

Evans, Melbourne, "Newton and the cause of gravity," *American Journal of Physics* 26 (1958), 619–24.

Feingold, Mordechai, *The Newtonian Moment: Isaac Newton and the Making of Modern Culture*, New York and Oxford: New York Public Library and Oxford University Press, 2004.

Feynman, Richard, *The Feynman Lectures on Physics*, Boston: Addison-Wesley, 1961.

Franks, Paul, *All or Nothing: Systematicity, Transcendental Arguments, and Skepticism in German Idealism*, Cambridge, MA: Harvard University Press, 2005.

Friedman, Michael, "Kant and Newton: Why Gravity is Essential to Matter," in Phillip Bricker and R. I. G. Hughes (eds.), *Philosophical Perspectives on Newtonian Science*, Cambridge, MA: MIT Press, 1990.

"Kant on Science and Experience," in Christia Mercer and Eileen O'Neill (eds.), *Early Modern Philosophy: Mind, Matter, and Metaphysics*, Oxford: Oxford University Press, 2005.

"Metaphysical Foundations of Natural Science," in Graham Bird (ed.), *A Companion to Kant*, Oxford: Blackwell, 2006.

"Newton and Kant on Absolute Space: From Theology to Transcendental Philosophy," in M. Bitbol, P. Kerszberg, and J. Petitot (eds.), *Constituting Objectivity: Transcendental Approaches to Modern Physics*, Berlin: Springer, forthcoming.

Kant's Construction of Nature, unpublished manuscript.

Funkenstein, Amos, *Theology and the Scientific Imagination from the Middle Ages to the Seventeenth Century*, Princeton: Princeton University Press, 1986.

Gabbey, Alan, "Force and Inertia in the Seventeenth Century: Descartes and Newton," in Stephen Gavkroger (ed.), *Descartes: Philosophy, Mathematics and Physics*, Sussex: Harvester Press, 1980.

"Philosophia Cartesiana Triumphata: Henry More (1646–1671)," in Thomas Lennon *et al.* (eds.), *Problems of Cartesianism*, Kingston and Montreal: McGill-Queen's University Press, 1982.

"Newton, Active Powers, and the Mechanical Philosophy," in I. Bernard Cohen and George Smith (eds.), *The Cambridge Companion to Newton*, Cambridge: Cambridge University Press, 2002.

Gagnebin, Bernard, "De la cause de la Pesanteur. Mémoire de Nicolas Fatio de Duillier," *Notes and Records of the Royal Society of London* 6 (1949), 105–24.

Garber, Daniel, *Descartes' Metaphysical Physics*, Chicago: University of Chicago Press, 1992.

"Descartes, Mechanics, and the Mechanical Philosophy," in Peter French, Howard Wettstein, and Bruce Silver (eds.), *Midwest Studies in Philosophy*, vol. XXVI, Boston: Blackwell, 2002.

"Physics and Foundations," in Katharine Park and Lorraine Daston (eds.), *The Cambridge History of Science,* vol. III: *Early Modern Science*, Cambridge: Cambridge University Press, 2006.

Goclenius, Rodolphus, *Lexicon Philosophicum*, Frankfurt, 1613.

Grant, Edward, "God and Natural Philosophy: The Late Middle Ages and Sir Isaac Newton," *Early Science and Medicine* 5 (2000), 279–98.

A History of Natural Philosophy, Cambridge: Cambridge University Press, 2007.

s'Gravesande, W. James, *Mathematical Elements of Natural Philosophy*, trans. by J. T. Desaguliers, London: W. Innys *et al.*, 1747.

Guicciardini, Niccolò, *Reading the Principia: The Debate on Newton's Mathematical Methods for Natural Philosophy from 1687 to 1736*, Cambridge: Cambridge University Press, 1999.

Hall, A. Rupert, "Newton and the Absolutes," in P. M. Harman and Alan Shapiro (eds.), *The Investigation of Difficult Things: Essays on Newton and the History of the Exact Sciences*, Cambridge: Cambridge University Press, 1992.

Hanson, Norwood Russell, *Patterns of Discovery: An Inquiry into the Conceptual Foundations of Science*, Cambridge: Cambridge University Press, 1958.

Perception and Discovery: An Introduction to Scientific Inquiry, San Francisco: Freeman, Cooper and Company, 1969.

"Hypotheses Fingo," in Robert Butts and John Davis (eds.), *The Methodological Heritage of Newton*, Toronto: University of Toronto Press, 1970.

Harman, P. M, *Metaphysics and Natural Philosophy: The Problem of Substance in Classical Physics*, Sussex: Harvester Press, 1982.

Harper, William, "Reasoning from Phenomena: Newton's Argument for Universal Gravitation and the Practice of Science," in Paul Theerman and Adele Seeff (eds.), *Action and Reaction: Proceedings of a Symposium to Commemorate the Tercentenary of Newton's Principia*, London: Associated University Presses, 1993.

Harper, William and George Smith, "Newton's New Way of Inquiry," in Jarrett Leplin (ed.), *The Creation of Ideas in Physics*, Dordrecht: Kluwer, 1995.

Harré, Rom, *Matter and Method*, New York: Macmillan, 1964.

Harrison, John, *The Library of Isaac Newton*, Cambridge: Cambridge University Press, 1978.

Hatfield, Gary, "Metaphysics and the New Science," in David Lindberg and Robert Westman (eds.), *Reappraisals of the Scientific Revolution*, Cambridge: Cambridge University Press, 1990.

"Was the Scientific Revolution a Revolution in Science?" in F. Jamil Ragep and Sally Ragep (eds.), *Tradition, Transmission, Transformation*, Leiden: E. J. Brill, 1996.

Descartes and the "Meditations," London: Routledge, 2003.

Heilbron, J. L., *Elements of Early Modern Physics*, Berkeley: University of California Press, 1982.

Henry, John, "'Pray do not ascribe that notion to me': God and Newton's Gravity," in James Force and Richard Popkin (eds.), *The Books of Nature and Scripture*, Dordrecht: Kluwer, 1994.

Herivel, John, *The Background to Newton's Principia: A Study of Newton's Dynamical Researches in the Years 1664–1684*, Oxford: Clarendon Press, 1965.

Hesse, Mary, *Forces and Fields: The Concept of Action at a Distance in the History of Physics*, London: Nelson, 1961.

"Comment on Howard Stein's 'On the Notion of Field in Newton, Maxwell, and Beyond,'" in Roger Stuewer (ed.), *Minnesota Studies in the Philosophy of Science*, vol. V, Minneapolis: University of Minnesota Press, 1970.

Hiscock, W. G. (ed.) *David Gregory, Isaac Newton, and their Circle: Extracts from David Gregory's Memoranda, 1677–1708*, Oxford: Oxford University Press, 1937.

Holton, Gerald, *Introduction to Concepts and Theories in Physical Science*, Cambridge, MA: Addison-Wesley, 1952.

Hutchison, Keith, "What Happened to Occult Qualities in the Scientific Revolution?" *Isis* 73 (1982), 233–53.

Huygens, Christiaan, "An Extract of a letter lately written by an ingenious person from Paris," *Philosophical Transactions of the Royal Society* (no. 96), 21 July 1673, 6086–7.

Œuvres complètes, ed. Johan Adriaan Vollgraff, The Hague: Nijhoff, 1888–1950.

The Pendulum Clock, or geometrical demonstrations concerning the motion of pendula as applied to clocks, trans. Richard J. Blackwell, Ames: Iowa State University Press, 1986.

Discours sur la cause de la Pesanteur (1690), in *Œuvres complètes*, vol. XXI; annotated and trans. Karen Bailey and George Smith as *Discourse on the Cause of Gravity* (unpublished manuscript).

Jacob, Margaret, "Christianity and the Newtonian Worldview," in David Lindberg and Ronald Numbers (eds.), *God and Nature: Historical Essays on the Encounter between Christianity and Science*, Berkeley: University of California Press, 1986.

Jammer, Max, *Concepts of Mass in Classical and Modern Physics*, Cambridge, MA: Harvard University Press, 1961.

Concepts of Force: A Study in the Foundations of Dynamics, New York: Harper, 1962.

Concepts of Mass in Contemporary Physics and Philosophy, Princeton: Princeton University Press, 2000.

Janiak, Andrew, "Space, Atoms and Mathematical Divisibility in Newton," *Studies in History and Philosophy of Science* 30 (2000), 203–30.

"Newton's Philosophy," in Edward Zalta (editor-in-chief), *Stanford Encyclopedia of Philosophy*, fall 2006 edition (http://plato.stanford.edu/entries/newton-philosophy/).

"Newton and the Reality of Force," *Journal of the History of Philosophy* 45 (2007), 127–47.

Joy, Lynn, "Scientific Explanation from Formal Causes to Laws of Nature," in Katharine Park and Lorraine Daston (eds.), *The Cambridge History of Science,* vol. III: *Early Modern Science*, Cambridge: Cambridge University Press, 2006.

Kant, Immanuel, *Kants gesammelte Schriften*, ed. Königlich preussischen Akademie der Wissenschaften, Berlin: Reimer, 1910–.

Koyré, Alexandre, *From the Closed World to the Infinite Universe*, Baltimore: Johns Hopkins University Press, 1957.

Newtonian Studies, Chicago: University of Chicago Press, 1968.

Lange, Marc, *An Introduction to the Philosophy of Physics: Locality, Fields, Energy, and Mass*, Oxford: Blackwell, 2002.

Leibniz, Gottfried Wilhelm, "A letter of M. Leibnitz to M. Hartsoeker," *Memoirs of Literature* 3 (1722 – reprint of 1712 edition), 453–60.

"A second letter of M. Leibnitz to M. Hartsoeker, dated July 12, 1711," *Memoirs of Literature* 5 (1722 – reprint of 1712 edition), 62–5.

Die mathematische Schriften, ed. C. Gerhardt, Berlin: A. Asher, 1849–.

Philosophical Essays, trans. Roger Ariew and Daniel Garber, Indianapolis: Hackett, 1989.

Die philosophischen Schriften, ed. C. Gerhardt, Berlin: Weidmann, 1890.

Locke, John, *The Works of John Locke, in nine volumes*, 9th edn, London, 1794.

An Essay Concerning Human Understanding, ed. Peter Nidditch, Oxford: Clarendon Press, 1975.

LoLordo, Antonia, *Pitrre Gassendi and the Birth of Early Modern Philosophy*, Cambridge: Cambridge University Press, 2007.

McCann, Edwin, "Lockean Mechanism," in A. J. Holland (ed.), *Philosophy, Its History and Historiography*, Dordrecht: Reidel, 1985.

McGuire, J. E., "Newton's 'Principles of Philosophy': An Intended Preface for the 1704 *Opticks* and a Related Draft Fragment," *British Journal for the History of Science* 5 (1970), 178–86.

"Atoms and the 'Analogy of Nature': Newton's Third Rule of Philosophizing," *Studies in History and Philosophy of Science* 1 (1970), 3–58.

"Boyle's Conception of Nature," *Journal of the History of Ideas* 33 (1972), 523–42.

"Newton on Place, Time and God: An Unpublished Source," *British Journal for the History of Science* 11 (1978), 114–29.

"Predicates of Pure Existence: Newton on God's Space and Time," in Phillip Bricker and R. I. G. Hughes (eds.), *Philosophical Perspectives on Newtonian Science*, Cambridge, MA: MIT Press, 1990.

"Natural Motion and its Causes: Newton on the 'vis insita' of bodies," in Mary Louise Gill and James Lennox (eds.), *Self-motion: From Aristotle to Newton*, Princeton: Princeton University Press, 1994.

Tradition and Innovation: Newton's Metaphysics of Nature, Dordrecht: Kluwer, 1995.

McGuire, J. E. and P. M. Rattansi, "Newton and the 'Pipes of Pan,'" *Notes and Records of the Royal Society* 21 (1966), 108–43.

McMullin, Ernan, *Newton on Matter and Activity*, Notre Dame: University of Notre Dame Press, 1978.

"The Significance of Newton's *Principia* for Empiricism," in Margaret Osler and Paul Lawrence Farber (eds.), *Religion, Science and Worldview: Essays in Honor of Richard Westfall*, Cambridge: Cambridge University Press, 1985.

"The Explanation of Distant Action: Historical Notes," in James Cushing and Ernan McMullin (eds.), *Philosophical Consequences of Quantum Theory: Reflections on Bell's Theory*, Notre Dame: University of Notre Dame Press, 1989.

"The Impact of Newton's *Principia* on the Philosophy of Science," *Philosophy of Science* 68 (2001), 279–310.

"The Origins of the Field Concept in Physics," *Physics in Perspective* 4 (2002), 13–39.

Maclaurin, Colin, *An Account of Sir Isaac Newton's Philosophical Discoveries in Four Books*, London, 1748.

Mandelbaum, Maurice, *Philosophy, Science and Sense Perception*, Baltimore: Johns Hopkins University Press, 1964.

Mandelbrote, Scott, "Isaac Newton and Thomas Burnet: Biblical Criticism and the Crisis of Late Seventeenth-century England," in James Force and Richard Popkin (eds.), *The Books of Nature and Scripture*, Dordrecht: Kluwer, 1994.

Maxwell, James Clerk, *The Scientific Papers of James Clerk Maxwell*, ed. W. D. Niven, Cambridge: Cambridge University Press, 1890.

Metzger, Hélène, *Attraction universelle et religion naturelle chez quelques commentateurs anglais de Newton*, Paris: Hermann et cie, 1938.

More, Henry, *An Antidote Against Atheism*, London: J. Flesher, 1655.

Immortality of the Soul, so farre forth as it is demonstrable from the knowledge of nature and the light of reason, London: J. Flesher, 1659.

A Collection of Several Philosophical Writings, London: J. Flesher, 1662.

Nagel, Ernest, *The Structure of Science: Problems in the Logic of Scientific Explanation*, New York: Harcourt, Brace & World, 1961.

Nelson, Alan, "Micro-Chaos and Idealization in Cartesian Physics," *Philosophical Studies* 77 (1995), 377–91.

Newton, Isaac, *Questiones Quædam Philosophicæ*, University Library, Cambridge, MSS Add. 3996; copy in Newton Papers of Indiana University, 1664.

De Gravitatione et Aequipondio Fluidorum, University Library, Cambridge, MSS Add. 4003; copy in Newton Papers of Indiana University, date unknown.

"A Letter of Mr. Isaac Newton, Professor of the Mathematicks in the University of Cambridge; containing his New Theory about Light and Colors," *Philosophical Transactions of the Royal Society* 6 (February 1672), 3075–87.

Philosophiae Naturalis Principia Mathematica, London: Royal Society, 1687.

Opticks, Or A Treatise of the Reflections, Refractions, Inflections & Colours of Light, 2nd English edn, London: Printers to the Royal Society, 1717.

The Mathematical Principles of Natural Philosophy, trans. Andrew Motte, London: Benjamin Motte, 1729.

De Mundi Systemate, London: J. Tonson *et al.*, 1731.

A Treatise of the System of the World, 2nd edn, trans. unknown, London: F. Fayram, 1731.

Principes mathématiques de la philosophie naturelle, trans. Madame la Marquise du Chatelet, with *Exposition Abregége du Systeme du Monde, et explication des principaux phénomenes astronomiques tirée des principes de M. Newton*, by Châtelet and Clairaut, Paris: Desaint & Saillant, 1749.

Four Letters from Sir Isaac Newton to Doctor Bentley, London: R. and J. Dodsley, 1756.

Isaaci Newtoni Opera quae exstant omnia, ed. Samuel Horsley, London: J. Nichols, 1779–85.

Opticks, Or A Treatise of the Reflections, Refractions, Inflections & Colours of Light, based on the fourth edition of 1730, New York: Dover, 1952.

Isaac Newton's Papers and Letters on Natural Philosophy, ed. I. Bernard Cohen and Robert Schofield, Cambridge, MA: Harvard University Press, 1958.

The Correspondence of Isaac Newton, ed. H. W. Turnbull *et al.*, Cambridge: Cambridge University Press, 1959–77.

Sir Isaac Newton's Mathematical Principles of Natural Philosophy and His System of the World, the Andrew Motte translation [1729] rev. and ed. Florian Cajori, Berkeley: University of California Press, 1960.

Unpublished Scientific Papers of Isaac Newton, ed. A. R. Hall and Marie Boas Hall, Cambridge: Cambridge University Press, 1962.

The Mathematical Papers of Isaac Newton, ed. D. T. Whiteside, with the assistance of M. A. Hoskin, Cambridge: Cambridge University Press, 1967–81.

Philosophiae Naturalis Principia Mathematica, ed. Alexandre Koyré and I. Bernard Cohen, with Anne Whitman, the 3rd edn with variant readings, Cambridge, MA: Harvard University Press, 1972.

Certain Philosophical Questions: Newton's Trinity Notebook, ed. J. E. McGuire and Martin Tamny, Cambridge: Cambridge University Press, 1983.

The Optical Papers of Isaac Newton, vol. I, ed. Alan Shapiro, Cambridge: Cambridge University Press, 1984.

The Principia: Mathematical Principles of Natural Philosophy, trans. I. Bernard Cohen and Anne Whitman, Berkeley: University of California Press, 1999.

Newton: Philosophical Writings, ed. Andrew Janiak, Cambridge: Cambridge University Press, 2004.

Normore, Calvin, "What is to be Done in the History of Philosophy," *Topoi* 25 (2006), 75–82.

Pampusch, Anita, *Isaac Newton's Notion of Scientific Explanation*, unpublished doctoral dissertation, University of Notre Dame, 1972.

Putnam, Hilary, *Reason, Truth, and History*, Cambridge: Cambridge University Press, 1981.

Rohault, Jacques, *Physica*, London: Jacobi Knapton, 1697 (includes Samuel Clarke's *Annotata*).

   System of natural philosophy: illustrated with Dr. Samuel Clarke's notes, taken mostly out of Sir Isaac Newton's Philosophy, trans. John Clarke, London: James Knapton, 1723.

Rosenberger, Ferdinand, *Isaac Newton und seine physikalischen Principien*, Leipzig: Barth, 1895.

Rutherford, Donald, "Innovation and Orthodoxy in Early Modern Philosophy," in Donald Rutherford (ed.), *The Cambridge Companion to Early Modern Philosophy*, Cambridge: Cambridge University Press, 2006.

Rynasiewicz, Robert, "By Their Properties, Causes and Effects: Newton's Scholium on Space, Time Place and Motion – I. The Text," *Studies in History and Philosophy of Science* 26 (1995), 133–53.

   "By Their Properties, Causes and Effects: Newton's Scholium on Space, Time, Place and Motion – II. The Context," *Studies in History and Philosophy of Science* 26 (1995), 295–321.

Sabra, A. I., *Theories of Light from Descartes to Newton*, 2nd edn, Cambridge: Cambridge University Press, 1981.

Sanford, David, "Locke, Leibniz and Wiggins on Being in the Same Place at the Same Time," *The Philosophical Review* 79 (1970), 75–82.

Schliesser, Eric and George Smith, "Huygens's 1688 Report to the Directors of the Dutch East Indian Company on the Measurement of Longitude at Sea and the Evidence it Offered Against Universal Gravity," *Archive for History of Exact Sciences* (forthcoming).

Sellars, Wilfrid, "Philosophy and the scientific image of man," in *Science, Perception, and Reality*, London: Routledge and Kegan Paul, 1963.

Shapiro, Alan, *Fits, Passions and Paroxysms: Physics, Method, and Chemistry and Newton's Theories of Colored Bodies and Fits of Easy Reflection*, Cambridge: Cambridge University Press, 1993.

Sklar, Larry, "Physics, Metaphysics and Method in Newton's Dynamics," in Richard Gale (ed.), *The Blackwell Guide to Metaphysics*, Oxford: Blackwell, 2002.

Smith, George, *Lecture Notes on the Newtonian Revolution*, Medford, MA: Department of Philosophy, Tufts University, 2000.

   "Comment on Ernan McMullin's 'The Impact of Newton's *Principia* on the Philosophy of Science,'" *Philosophy of Science* 68 (2001), 327–38.

   "The Methodology of the *Principia*," in I. Bernard Cohen and George Smith (eds.), *The Cambridge Companion to Newton*, Cambridge: Cambridge University Press, 2002.

   "From the Phenomenon of the Ellipse to an Inverse-Square Force: Why Not?" in David Malament (ed.), *Reading Natural Philosophy: Essays in the History and Philosophy of Science and Mathematics*, Chicago: Open Court Press, 2002.

Snobelen, Stephen, "'The True Frame of Nature': Isaac Newton, Heresy, and the Reformation of Natural Philosophy," in John Brooke and Ian Maclean (eds.), *Heterodoxy in Early Modern Science and Religion*, Oxford: Oxford University Press, 2005.

Stein, Howard, "Newtonian Space–Time," in Robert Palter (ed.), *The Annus Mirabilis of Sir Isaac Newton, 1666–1966*, Cambridge, MA: MIT Press, 1970.

"On the Notion of Field in Newton, Maxwell, and Beyond," in Roger Stuewer (ed.), *Minnesota Studies in the Philosophy of Science*, vol. V, Minneapolis: University of Minnesota Press, 1970.

"On Locke, 'the Great Huygenius, and the incomparable Mr. Newton,'" in Phillip Bricker and R. I. G. Hughes (eds.), *Philosophical Perspectives on Newtonian Science*, Cambridge, MA: MIT Press, 1990.

"'From the Phenomena of Motions to the Forces of Nature': Hypothesis or Deduction?" *Philosophy of Science Association* 2 (1990), 209–22.

"On Philosophy and Natural Philosophy in the Seventeenth Century," *Midwest Studies in Philosophy*, vol. XVIII, Minneapolis: University of Minnesota Press, 1993.

"Newton's Metaphysics," in I. Bernard Cohen and George Smith (eds.), *The Cambridge Companion to Newton*, Cambridge: Cambridge University Press, 2002.

Steinle, Friedrich, *Newtons Entwurf "Über die Gravitation." Ein stück entwicklungsgeschichte seiner mechanik*, Stuttgart: Franz Steiner Verlag, 1991.

Suppes, Patrick, "Descartes and the Problem of Action at a Distance," *Journal of the History of Ideas* 15 (1954), 146–52.

Voltaire, *Eléments de la philosophie de Newton*, Paris, 1738.

Warren, Daniel, "Kant's Dynamics," in Eric Watkins (ed.), *Kant and the Sciences*, Oxford: Oxford University Press, 2001.

Force in Newton's Physics, New York: Wiley, 1971.

Never at Rest, Cambridge: Cambridge University Press, 1980.

Wilson, Margaret, *Ideas and Mechanism*, Princeton: Princeton University Press, 1999.

Woolhouse, Roger, "Locke and the Nature of Matter," in Christia Mercer and Eileen O'Neill (eds.), *Early Modern Philosophy: Mind, Matter and Metaphysics*, Oxford: Oxford University Press, 2005.

Index